ORDER NUMBER EA-338

FLIGHT THEORY FOR PILOTS

THIRD EDITION

By Charles E. Dole
Institute of Safety and Systems Management
University of Southern California

International Standard Book Number 0-89100-338-X
For sale by: IAP, Inc., A Hawks Industries Company
Mail To: P.O. Box 10000, Casper, WY 82602-1000
Ship To: 7383 6WN Road, Casper, WY 82604-1835
(800) 443-9250 ❖(307) 266-3838 ❖FAX: (307) 472-5106
HBC0791 Printed in the USA

PREFACE TO FIRST EDITION

"It is quite possible to fly without knowing the first thing about aerodynamics (some of us have been doing it for years)."

This anonymous statement is certainly true, but it does not go far enough. It should continue, "however it is reasonable to assume that such knowledge will help a pilot to have a better understanding of the capabilities and limitations of his aircraft."

This book is written for the typical pilot, who has no engineering background and no desire to become involved with mathematical solutions, but who still would like to expand his knowledge of flight theory. From twenty one years of teaching this subject to experienced pilots of the U.S. Air Force, Army, Navy, Marines, Coast Guard and many foreign countries, I would estimate that this category would include at least 75% of these students.

The book does not attempt to transform pilots into engineers, nor does it try to teach a pilot how to design an airplane or fly one. What it does try to explain, as simply as possible, are the basics of aerodynamics as they apply to flying an airplane. In some cases formulas are derived or presented, but they are simplified to a great extent and a real insight into the subject can be obtained even if they are ignored completely.

A noted educational psychologist, Lee J. Cronbach, said, "Greater understanding produces more rapid learning, better retention, and better adaptation to new conditions." Thus, in this book, I hope to substitute the understanding method for the "do it this way because I say so" method of learning that is so often used in flight training.

This book is the text used in the United States Air Force Flying Safety Officer course presented by the University of Southern California.

I am grateful for the assistance of my fellow staff members, especially Mr. Gay Jones, Dr. Harold Roland, Mr. Richard Davis, and the students' of the USAF Flight Safety Officer course for suggestions and criticism of the manuscript. Previous work done in this field by Professor H. H. Hurt Jr., of the University of Southern California, in his book, AERODYNAMICS FOR NAVAL AVIATORS, together with many illustrations from that text, that I have reproduced in this book are gratefully acknowledged.

Charles E. Dole, EdD

1503 Franklin Ave.
Redlands, CA 92373, USA
August 1984

PREFACE TO THE SECOND EDITION

Four major revisions to the first edition have been made. It was assumed that the reader of the book would not be interested in the aspects of high speed flight. This did not prove to be a valid assumption, so a chapter on this subject has been added. The second revision was to add mathematical problems to the existing non-mathematical problems at the end of each chapter. The third revision was to improve the quality of the dot matrix printing. An improved printer has been used in this edition. Finally the page size has been increased from 5 1/2 x 8 to 8 1/2 x 11 inches.

The text has now been accepted for use in the following USAF courses:

> Flight Safety Officer Course
> Aircraft Mishap Investigation Course
> International Flight Safety Officer Course
> Air National Guard Aircraft Mishap Prevention Course.

I am grateful to Captain David Hernandez USAF for proof reading the second edition text.

Charles E. Dole, EdD

1503 Franklin Ave. Redlands, CA 92373
January, 1987
2nd Printing, August 1987
3rd Printing, March 1988

PREFACE TO THE THIRD EDITION

The flying skills of the readers of the book appears to include many who are helicopter qualified. In response to many requests from this group, a third edition including a chapter on helicopter aerodynamics has been added.

More printing improvements have also been made, including the use of a scientific word processor.

I am grateful for the many Universities and Colleges that have accepted the book and text for non-engineering level courses.

The third edition is now being published by IAP, Inc. of Casper, Wyoming.

Charles E. Dole, EdD

1503 Franklin Ave. Redlands, CA 92373
May 1989

CONTENTS

Contents

Contents

Contents

CHAPTER ONE

INTRODUCTION

A basic understanding of the physical laws that affect aircraft in flight, and on the ground, is a prerequisite for the study of aerodynamics. A brief summary of the branch of physics called mechanics is, therefore, presented here. Concepts of forces, vectors, moments, Newton's laws, mass, work, energy, power, and friction are discussed in this chapter. For a more detailed explanation of these subjects, the reader should consult any high school physics text.

BASIC QUANTITIES

Because the metric system of measurement has not yet been widely accepted in America, the U.S. customary system of measurements is used in this book.

The fundamental units are:

Force . . . . pounds (lb.)
Distance . . . . feet (ft.)
Time . . . . seconds (sec.)

From the fundamental units, other quantities can be found.

Velocity (distance/time) . . . . feet per second (fps)
Area (distance squared) . . . . square feet (ft^2)
Pressure (force/unit area) . . . . lb/ft^2 (psf)
Acceleration(change in velocity) . . . . . . ft/sec/sec (fps^2)

Aircraft airspeed is measured in knots (nautical miles per hour) or in Mach number (the ratio of airspeed to the speed of sound). Rates of climb or descent are measured in feet per minute, so quantities other than those above are used in some cases.

FORCE

A force is a push or a pull tending to change the state of motion of a body.
To define a force we must know two things:

(1) The size of the force and,

(2) the direction of the force's action.

Introduction

Typical forces acting on an airplane in flight are shown in Figure 1.1.

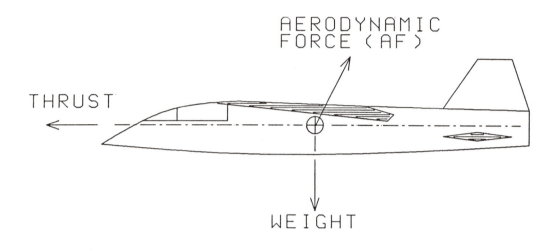

Figure 1.1   Forces on an airplane in steady flight.

There are three principal forces acting on the airplane.

These are *Thrust, Weight,* and *Aerodynamic Force* (AF).

Thrust is the force that moves the aircraft through the air.It is provided by the propeller or jet engine.

Weight is the force of gravity and it always acts toward the center of the earth.

The *Aerodynamic Force* is a result of air pressures acting on the airplane as it moves through the air.

To simplify the study of the aerodynamic force and its reaction on the airplane, we resolve the AF into two components.

The component that is 90 degrees to the flight path and acts towards the top of the airplane, we call *Lift.*

The component that is parallel to the flight path and acts toward the rear of the airplane, we call *Drag.*

Figure 1.2 shows typical forces acting on an airplane in flight after the Aerodynamic Force has been replaced by the Lift and Drag component vectors.

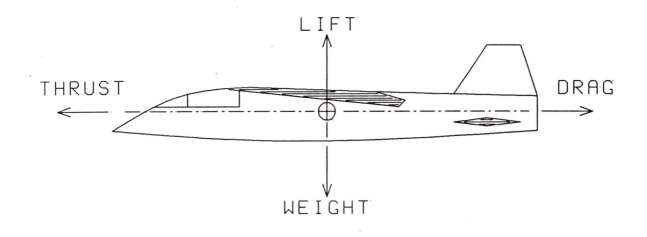

Figure 1.2   Resolved forces on an airplane in steady flight.

Lift, thrust, weight, and drag are all considered to act through the aircraft's *Center of Gravity*, CG. The CG can be thought of as the center of balance.

SCALAR AND VECTOR QUANTITIES

A quantity that has both size and direction is called a *vector quantity*, or more simply, a *vector*. Forces are vectors, as are accelerations, velocities, and displacements.

A quantity that has size only, on the other hand, is called a *scalar quantity*, or simply, a *scalar*. The quantities of mass, time, and temperature are examples of scalar quantities.

Distance is a scalar but, if we consider the direction of the distance, then it is called *displacement*, which is a vector.

If a car traveled 50 miles, the distance is a scalar. But if the car traveled 50 miles **to the north**, the car's displacement is a vector quantity.

Scalar Addition

Scalar quantities can be added (or subtracted) by simple arithmetic. For example, if you have 10 gallons of gas in your car's tank and you stop at a gas station and top off your tank with 8 gallons more, your tank now holds 18 gallons.

Vector Addition

Vector addition is more complicated than scalar addition. Vector quantities are conveniently represented by arrows. The length of the arrow represents the size of the quantity, and the orientation of the arrow represents the directional property of the quantity. For example if we consider the top of this page as representing north, and we want to show the velocity of an aircraft flying to the east at an airspeed of 300 knots, the velocity vector is as shown in Figure 1.3.

3

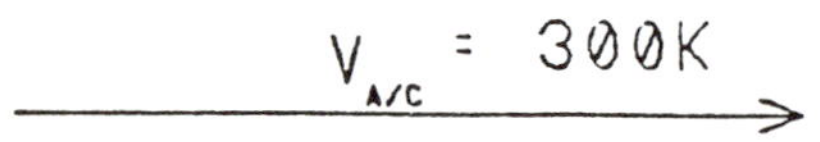

$$V_{A/C} = 300K \longrightarrow$$

Figure 1.3   Vector of an eastbound aircraft.

If the aircraft encounters a 30 knot wind from the north, the wind vector is as shown in Figure 1.4.

$$\downarrow V_H = 30K$$

Figure 1.4   Vector of a north wind.

To find the aircraft's track, groundspeed, and drift angle, we vectorially add these two vectors, as follows. Place the tail of the wind vector, V_W, at the arrow of the aircraft vector, $V_{A/C}$, and then draw a straight line from the tail of the aircraft vector to the arrow of the wind vector. The straight line that you drew is the *resultant vector*, V_R, and represents the path of the aircraft over the ground. The length of the resultant represents the groundspeed, and the angle between the aircraft vector and the resultant vector is the drift angle. This is shown in Figure 1.5.

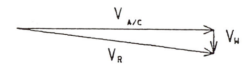

Figure 1.5   Vector addition

Vector Resolution

It is often desirable to replace a given vector by two or more other vectors. This is called *vector resolution*. The resulting vectors are called *component vectors* of the original vector and, if added vectorially, they will produce the original vector.

For example, if an aircraft is in a steady climb, at a constant airspeed and climb angle, the groundspeed and rate of climb can be found by vector resolution. The flight path and velocity can be shown by vector $V_{A/C}$ in Figure 1.6.

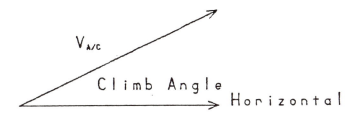

Figure 1.6 *Vector of an aircraft in a climb.*

To resolve vector, $V_{A/C}$, into a component, V_H, parallel to the horizontal, which represents the groundspeed, and into a vertical component, V_V, which represents the rate of climb, we simply draw a straight line vertically upward from the horizontal to the tip of the arrow, $V_{A/C}$. This vertical line represents the rate of climb, and the horizontal line represents the groundspeed of the aircraft. This is shown in Figure 1.7.

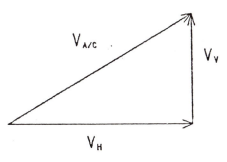

Figure 1.7 *Vectors of ground speed and rate of climb.*

MOMENTS

Up to now we have assumed that the forces on the airplane acted at its center of gravity. This is not always true.

Suppose that the pilot wants the airplane to climb. He can pull back on the control stick (or yoke). This causes the elevator to move upward, and a downward force on the tail surfaces results. This downward force pushes the tail down and the nose pitches up.

This is called a *pitching moment*. A *moment* has a turning or rotating effect.

Moments are caused by forces that do not act through the point of rotation. The value of the moment is equal to the force multiplied by its *moment arm*.

The moment arm is the shortest distance from the point of rotation to the line of action of the force. The units of moments are foot-pounds (ft-lb) or inch-pounds (in-lb).

The control stick (or yoke) and rudder pedals of an airplane are used to produce moments about the CG and thus change the airplane's attitude. We have seen that pulling on the stick produces a nose-up pitching moment. Likewise a push will produce a nose-down pitching moment.

Sideward movements of the stick activate the ailerons. Moving the stick to the right, for instance, moves the left aileron down and the right aileron up. Upward forces are developed on the left wing and downward forces are developed on the right wing and a *rolling moment* to the right is created.

In addition to the control stick, there is another control device. This is the rudder. By pushing on the rudder pedals, the rudder is displaced and a sideward force is produced. This sideward force acts behind the CG and a directional moment, called a *yawing moment*, results.

EQUILIBRIUM

An airplane is said to be in *equilibrium* when there are no unbalanced forces or unbalanced moments.

An engineer would say, "The sum of the forces = 0 and the sum of the moments = 0."

An airplane is in equilibrium when the airspeed is constant and there are no pitching, yawing, or rolling moments.

You will notice that the above sentence does not specify that the airplane must be at a constant altitude.

The airplane can be in a climb or glide and still be in equilibrium, provided that it is not changing its airspeed. We will discuss this in more detail in later chapters.

NEWTON'S LAWS

Sir Isaac Newton summarized three generalizations about force and motion. These are known as the Laws of Motion.

Newton's First Law

In simple language, the first law says: "A body at rest will remain at rest and a body in motion will remain in motion, in a straight line, unless it is acted upon by an unbalanced force."

This helps us understand why the airplane in Figure 1.2 will fly at a constant airspeed, if the thrust and drag are equal in value but opposite in direction. In this case there will be no unbalanced force in the direction of flight.

Likewise, if the lift and weight are equal in value and opposite in direction, there will be no change in altitude.

If we say that an object is at rest or if it is moving at a certain speed, we are describing its *state of motion.*

All objects have a resistance to a change in their state of motion. This property is called *inertia.*

Speed is the amount of distance that a moving object covers in a unit of time.

Velocity, on the other hand, is more complex. Velocity not only includes the speed of an object, but the direction of movement as well.

If we say that a car is moving at 40 miles per hour, we are describing its speed. However, if we say that a car is moving at 40 miles per hour **to the north,** we are describing its velocity.

All objects have inertia and so they resist having their velocity changed.

Newton said that an unbalanced force must be acting on an object, if its velocity is changing.

A change in velocity is called *acceleration* (a). The basic units of velocity are feet per second, (fps). The basic units of acceleration are feet per second per second, usually called feet per second squared, (fps^2).

Don't be confused by the "squared." You really don't square anything, it's just the unit's name.

Newton's Second Law

The second law states: "If a body is acted on by an unbalanced force, the body will accelerate in the direction of the force and the acceleration will be directly proportional to the force and inversely proportional to the mass of the body."

If the body had some velocity before it was acted upon by the unbalanced force and the force is in the same direction as the original velocity, the body will be speeded up. This is a positive (+) acceleration.

If the unbalanced force acts in the opposite direction to the original velocity, the body will be slowed. The acceleration will have a negative (–) value and is called *deceleration.*

If the unbalanced force acts so that the direction of the body is changed, but the speed is not changed, the acceleration is called *radial acceleration.*

Newton's second law tells us how the acceleration is affected by the unbalanced force and the *Mass*, m, of the object. Mass is discussed below.

The acceleration is directly proportional to the unbalanced force, F.

$$a \sim F$$

The acceleration is inversely proportional to the mass of the object.

$$a \sim \frac{1}{m}$$

Newton's second law can be written as an equation: $a = \frac{F}{m}$. You probably know it as:

$$F = ma \tag{1.1}$$

Newton's Third Law

The third law states: "For every action force there is an equal and opposite reaction force."
Note that for this law to have any meaning, there must be an an interaction between the force and the object. The recoil of a gun is a good example of this law.

MASS

Mass is a measure of the amount of material in a body.

Weight, on the other hand, is a force caused by the gravitational attraction of the earth, moon, sun, or other heavenly bodies. Weight will vary, depending upon where the body is located in space. Mass will not vary with location.

$$\text{Weight (W)} = \text{mass (m)} \times \text{acceleration of gravity (g)}$$

$$W = mg \text{ or } m = \frac{W}{g} \text{ (lb-sec}^2\text{/ft)} \tag{1.2}$$

The units of mass are very confusing, so we give the units a "nickname." We call them "slugs."

WORK

In physics, work has a meaning different from the popular definition. Work requires that a force must move an object in the direction of the force. Another way of saying this is: "only the component of the force in the direction of movement does any work." You can push against a solid wall until you are blue in the face but, unless the wall moves, you are not doing any work.

$$\text{Work} = \text{Force x Distance}$$

ENERGY

Energy is the ability to do work. There are many kinds of energy; solar, chemical, heat, nuclear, and others. One type of energy of particular interest to us in aviation is *mechanical energy*. There are two kinds of mechanical energy.

The first is called *potential energy of position*, or more simply called *potential energy*, PE. No movement is involved in calculating PE.

A good example of this kind energy is water stored behind a dam. If released, the water would be able to do work, such as running a generator. It has the potential for doing work. PE equals the weight of an object, W, multiplied by its height, h, above some base plane:

$$PE = Wh \text{ (ft-lb)} \tag{1.3}$$

The second kind of mechanical energy is called *kinetic energy*, KE. As the name implies, kinetic energy requires movement of an object. It is a function of the mass of the object, m, and its velocity, V.

$$KE = \frac{1}{2}mV^2 \text{ (ft-lb)} \tag{1.4}$$

The total mechanical energy, TE, of an object is the sum of its PE and KE:

$$TE = PE + KE \tag{1.5}$$

There is a very important principle of energy which states, "all the energy in the universe is conserved."
Another way of saying this is, "energy cannot be created nor destroyed, but can change in form."

Both potential energy and kinetic energy can change in value, but the total energy must remain the same.

A simple example of transferring energy is the roller coaster. The car and passengers are first moved up the ramp to the highest point on the ride. As the car is moving up the ramp it is gaining potential energy. As the car starts downward, this potential energy is being transformed into kinetic energy. As the car accelerates, the kinetic energy increases. When the car reaches ground level, all the PE has been transformed into KE, and the speed of the car is at a maximum. As the car starts up the next ramp, the KE is transformed back into PE, and the car slows down.

POWER

Note that in our discussion of work and energy we have not mentioned time. *Power* is "the rate of doing work".

$$Power = \frac{work}{time} = \frac{(force)\ (distance)}{time}$$

But

$$\frac{distance}{time} = speed.$$

So:

$$Power = (force)\ (speed)\ (ft\text{-}lb/sec)$$

James Watt defined the term horsepower as 550 ft-lb/sec.

$$Horsepower = \frac{(force)\ (speed)}{550}$$

If airspeed is measured in knots, V_K, rather than in feet per second, and the force is Thrust, the equation becomes:

$$HP = \frac{(Thrust)\ (V_k)}{325} = \frac{TV_k}{325} \tag{1.6}$$

Equation (1.6) is useful in comparing thrust-producing aircraft (turbojets) with power-producing aircraft (propeller airplanes and helicopters).

FRICTION

Friction is a force which resists movement between surfaces that are in contact with each other.

If we examined the surface of a material through a magnifying glass we would see that even the smoothest of surfaces is really very rough.

These rough surfaces tend to lock together and friction forces result when the surfaces are caused to slide past each other. This type of friction is called *sliding friction*.

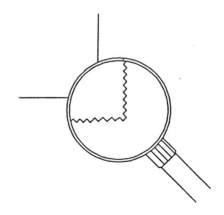

We also have friction between a round body, such as an airplane tire, rolling on a solid surface. This kind of friction is called *rolling friction*.

There is also friction when a solid body moves through a fluid. An airplane passing through air is resisted by friction and *skin friction drag* results.

10

All friction is not undesirable. We try to reduce the effects of friction in machines, but if you are trying to stop an airplane upon landing, you will want high friction forces.

There are several factors that determine how much stopping (braking) friction can be developed. It has been found that the braking force is directly proportional to the squeezing force between the braked tires and the runway. This is called the *normal force*

For an automobile with four wheel brakes, the normal force is the weight of the car. Most airplanes, however, have brakes on the main wheels only, so the normal force is only that amount of the weight that rests on them.

$$\frac{\text{friction force}}{\text{normal force}} = \text{constant}$$

The constant that relates the friction forces to the normal forces is called the *coefficient of friction*. We use the Greek letter Mu, μ, for this coefficient.

We use a lot of "coefficients" in aerodynamics. A coefficient is a dimensionless number that is constant for a certain set of conditions.

Besides the normal force on the braked wheels, there are other factors that influence the coefficient of friction. Among these are; runway surface material and condition (dry, wet, icy, etc.), tire composition and tread, inflation pressure, and the amount of brake pressure applied.

If too much brake pressure is applied the brakes will lock and the wheels will stop rotating. When this happens the tires skid and rubber is melted and burned from the tread. In addition to the possibility of causing a blow-out, the coefficient of friction is substantially reduced, as shown in Figure 1.8, and braking distances are increased.

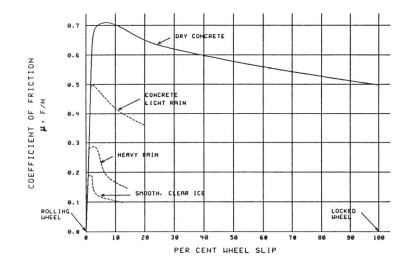

Figure 1.8 Coefficients of friction.

11

Introduction

The relationship between braking forces, normal forces, and the coefficient of friction can be written by a simple equation:

$$F_B = \mu N \qquad (1.7)$$

where
    F_B = braking force, (lb)
    μ = coefficient of friction (dimensionless)
    N = normal force

SYMBOLS AND UNITS

English Symbols

a	Acceleration (ft/sec^2)
AF	Aerodynamic Force (lb)
KE	Kinetic Energy (ft-lb)
PE	Potential Energy
TE	Total Energy
F	Force (lb)
F_B	Braking Force
g	Acceleration of gravity = 32 fps^2 (on earth)
h	Height (ft)
HP	Horsepower
m	Mass (slugs,lb-sec^2/ft)
N	Normal force (lb)
P	Pressure (lb/ft^2)
V	Speed (ft/sec)
V_K	Airspeed (knots)
W	Weight (lb)

Greek Symbol

μ (mu)   Coefficient of Friction (dimensionless)

EQUATIONS

1.1 $F = ma$

1.2 $W = mg$

1.3 $PE = Wh$

1.4 $KE = \frac{1}{2} mV^2$

1.5 $TE = PE + KE$

1.6 $HP = \frac{TV_K}{325}$

1.7 $F_B = \mu N$

PROBLEMS

1. Lift and Drag forces on an airplane are:

    a. scalars and are resultants of the Aerodynamic Force.
    b. vectors and are resultants of the Aerodynamic Force.
    c. scalars and are components of the Aerodynamic Force.
    d. vectors and are components of the Aerodynamic Force.

2. Lift on an airplane acts:

    a. parallel to the flight path.
    b. opposite to the weight.
    c. perpendicular to the flight path.
    d. perpendicular to the horizontal.

3. Which of the following are all vector quantities?

    a. Velocity, time, displacement.
    b. Displacement, acceleration, force.
    c. Force, displacement, volume.
    d. Velocity, mass, force.

4. An airplane is flying to the south at 200 knots and encounters a 20 knot wind from the north. The groundspeed of the airplane will be:

    a. 180 knots with no drift.
    b. 200 knots with a drift to the east.
    c. 220 knots with no drift.
    d. 200 knots with a drift to the west.

5. An airplane is making a constant airspeed, level turn. Its velocity:

    a. is increasing.
    b. is decreasing.
    c. remains the same.
    d. is changing.

6. Which dimension of the drawing is the moment arm?

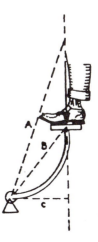

    a.
    b.
    c.
    d. None of the above.

7. An airplane is making its takeoff run. It is:

    a. In equilibrium.
    b. Accelerating according to Newton's third law.
    c. Not in equilibrium.
    d. About to crash because F = ma.

8. The term "work" means:

    a. A body moves as a result of balanced forces.
    b. A frictionless body moves at constant velocity.
    c. A force is exerted upon a body.
    d. A body moves as a result of an unbalanced force.

9. How much horsepower does a 10,000 pound thrust jet engine produce?

    a. 10,000 HP.
    b. 30.77 HP.
    c. Zero HP.
    d. Can't calculate until velocity is known.

10. How much greater will the stopping force be with the brakes applied for 10% slip as with the brakes locked (100% slip)? (see Figure 1.8)

    a. Same.
    b. 7/5.
    c. Twice.
    d. Need to know the weight.

11. An airplane weighs 16,000 lb. The local gravitational acceleration, g, is 32 fps^2. What is the mass of the airplane?

12. The above airplane accelerates down the runway with a net force of 6,000 lb. Find the acceleration of the airplane.

13. An airplane is towing a glider to altitude. The tow rope is 20° below the horizontal and has a tension force of 300 lb. exerted on it by the airplane.

    Find the horizontal drag of the glider and the amount of the lift that the rope is providing to the glider. Sin 20° = 0.342; cos 20° = 0.940.

14. A jet airplane is climbing at a constant airspeed in no-wind conditions. The plane is directly over a point on ground that is 4 statute miles (21,120 ft.) from the takeoff point and the altimeter reads 15,840 ft. Find the tangent of the climb angle and the distance that it has flown through the air.

15. Find the distance, s, and the force, F, on the seesaw fulcrum shown in the figure. Assume that the system is in equilibrium.

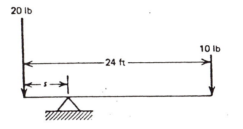

16. An airplane weighs 16,000 lb. and is flying at 5000 ft. altitude and at an airspeed of 200 fps.
    Find:
    (a) the potential energy,
    (b) the kinetic energy,
    (c) the total energy.

If the pilot dives the plane until the airspeed is 400 fps, what is the altitude? (assume no drag increase as speed increases.)

17. An airplane's turbojet engine produces 10,000 lb. of thrust at 162.5 knots true airspeed. What is the equivalent horse power that it is producing?

18. An aircraft weighs 2400 lb. and has 75% of its weight on the main (braking) wheels. If the coefficient of friction is 0.7, find the braking force, F_B, on the airplane.

19. A 100 pound weight is resting on a frictionless ramp and is restrained by a rope as shown in the figure. What is the tension force in the rope?

Sin 30° = 0.5

Cos 30° = 0.866

Tan 30° = 0.577

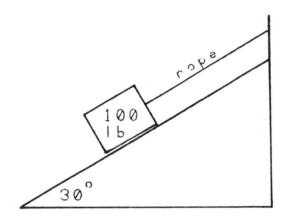

20. Now assume that there is friction present in the above figure and that the coefficient of friction is 0.5. What is the tension force in the rope?

CHAPTER TWO

AIR FLOW AND AIRSPEED MEASUREMENT

PROPERTIES OF THE ATMOSPHERE

The aerodynamic forces and moments acting on an aircraft in flight are due, in great part, to the properties of the air mass in which the aircraft is flying. The atmosphere is composed of approximately (by volume) 78 percent nitrogen, 21 percent oxygen, and 1 percent other gases. The most important properties of air that affect aerodynamic behavior are its static pressure, temperature, density, and viscosity.

Static Pressure

The *static pressure*, P, of the air is simply the weight per unit area of the air above the level under consideration. For instance, the weight of a column of air, with a cross sectional area of one square foot and extending upward from sea level through the atmosphere, is 2116 lb. The sea level static pressure is, therefore, 2116 psf (or 14.7 psi).

Static pressure is reduced as altitude is increased because there is less air weight above. At 18,000 ft altitude, the static pressure is about half that at sea level.

Another commonly used measure of static pressure is *inches of mercury*. On a standard sea level day the air's static pressure will support a column of mercury (Hg) that is 29.92 inches high.

Weather reports use a third method of measuring static pressure called *millibars*. Sea level standard is 1013 millibars.

In addition to these rather confusing systems, there are the metric measurements in use throughout most of the world.

In aerodynamics it is convenient to use pressure ratios, rather than actual pressures. Thus the units of measurement are canceled out. The Greek symbol for pressure ratio is δ (delta).

$$\delta = \frac{P}{P_0}$$

(2.1)

where P_0 is the sea level standard static pressure (2116 psf or 29.92 in.Hg).

Thus a pressure ratio of 0.5 means that the ambient pressure is one-half of the standard sea level value.

At 18,000 ft, on a standard day, the pressure ratio is 0.4992.

Temperature

The commonly used measures of temperature are Fahrenheit, F, and Celsius, C, (formerly called centigrade) scales. Neither degrees F or C are based upon absolute zero and can not be used in calculations. Absolute temperature must be used instead. Absolute zero in the Fahrenheit system is –460° and in the Celsius system is –273°. To convert from the Fahrenheit system to the absolute system, called Rankine, R, add 460 to the F°. To convert from the Celsius system to the absolute system, called Kelvin, K, add 273 to the C°.

The symbol for absolute temperature is, T, and the symbol for sea level standard temperature is T_o.

In the Rankine System, T_o = 519° (59°F)
In the Kelvin System, T_o = 288° (15°C)

By using temperature ratios rather than actual temperatures, the units cancel. The Greek symbol for temperature ratio is θ (theta).

$$\theta = \frac{T}{T_o} \quad (Tabs)$$

(2.2)

At sea level, on a standard day, θ_o = 1.0. Temperature decreases with altitude until the tropopause is reached (36,089 ft. on a standard day); then it remains constant until about 85,000 ft. The temperature at the tropopause is –69.7 °F and θ = 0.7519.

Density

The density of air is its most important property in the study of aerodynamics. Density is defined as the mass of the air per unit volume. The Greek symbol for density is ρ (rho).

$$\rho = \frac{mass}{unit\ volume} \quad (slugs/ft^3)$$

Standard sea level density, ρ_o = 0.002377 slugs/ft³. Density decreases with altitude. At about 22,000 ft, the density is about one half of sea level density. Actually it is 0.001183 slugs/ft³.
Once again, it is desirable in aerodynamics to use density ratios, rather than the actual values of density. The symbol for density ratio is σ(sigma).

$$\sigma = \frac{\rho}{\rho_o}$$

(2.3)

Density is directly proportional to pressure and inversely proportional to absolute temperature, as shown by the universal gas law:

$$\rho = \frac{P}{RT}$$

(2.4)

and so:

$$\frac{\rho}{\rho_o} = \frac{P/RT}{P_o/RT_o} = \frac{P/P_o}{T/T_o}$$

where R is the gas constant. Therefore:

$$\sigma = \frac{\delta}{\theta} \qquad\qquad (2.5)$$

Viscosity

Viscosity can be simply defined as the internal friction of a fluid caused by molecular attraction which makes it resist its tendency to flow.

The viscosity of the air is important when discussing airflow in the region very close to the surface of an aircraft. This region is called the *boundary layer*. We will discuss viscosity in more detail when we take up the subject of boundary layer theory.

ICAO STANDARD ATMOSPHERE

To provide a basis for comparing aircraft performance at different parts of the world and under varying atmospheric conditions, the performance data must be reduced to a set of standard conditions. These have been defined by the International Civil Aviation Organization (ICAO) and are compiled in a standard atmosphere table.

An abbreviated table is shown here as Table 2.1. Columns in the table show standard day density, density ratio, pressure, pressure ratio, temperature, temperature ratio, and speed of sound at various altitudes.

Two types of altitude interest the pilot: pressure altitude and density altitude.

Pressure altitude is that altitude in the standard atmosphere corresponding to a certain static pressure. For instance, if the pressure at a certain altitude is 1455 psf, the pressure ratio, $\delta = 1455/2116 = 0.6877$. Entering Table 2.1 with this value, we find a corresponding pressure altitude of 10,000 ft.

Density altitude is found by correcting pressure altitude for nonstandard temperature conditions. If a density ratio of 0.6292 is calculated, the density ratio column in Table 2.1 reveals that this value corresponds to a density altitude of 15,000 ft.

Density altitude affects aircraft performance. Therefore performance charts are drawn up for various density altitudes.

Airflow and Airspeed Measurement

Table 2.1 Standard Atmosphere Table

ATLTITUDE FT	DENSITY RATIO σ	√σ	PRESSURE RATIO δ	TEMPER- ATURE °F	TEMPER- ATURE RATIO θ	SPEED OF SOUND a KNOTS	KINEMTAIC VISCOSITY ν FT²/SEC
0	1.0000	1.0000	1.0000	59.00	1.0000	661.7	.000158
1000	0.9711	0.9854	0.9644	55.43	0.9931	659.5	.000161
2000	0.9428	0.9710	0.9298	51.87	0.9862	657.2	.000165
3000	0.9151	0.9566	0.8962	48.30	0.9794	654.9	.000169
4000	0.8881	0.9424	0.8637	44.74	0.9725	652.6	.000174
5000	0.8617	0.9283	0.8320	41.17	0.9656	650.3	.000178
6000	0.8359	0.9143	0.8014	37.60	0.9587	647.9	.000182
7000	0.8106	0.9004	0.7716	34.04	0.9519	645.6	.000187
8000	0.7860	0.8866	0.7428	30.47	0.9450	643.3	.000192
9000	0.7620	0.8729	0.7148	26.90	.09381	640.9	.000197
10000	0.7385	0.8593	0.6877	23.34	0.9312	638.6	.000202
15000	0.6292	0.7932	0.5643	5.51	0.8969	626.7	.000229
20000	0.5328	.07299	0.4595	-12.32	0.8625	614.6	.000262
25000	0.4481	0.6694	0.3711	-30.15	0.8281	602.2	.000302
30000	0.3741	0.6117	0.2970	-47.98	0.7937	589.5	.000349
35000	0.3099	0.5567	0.2353	-65.82	0.7594	576.6	.000405
*36089	0.2971	0.5450	0.2234	-69.70	0.7519	573.8	.000419
40000	0.2462	0.4962	0.1851	-69.70	0.7519	573.8	.000506
45000	0.1936	0.4400	0.1455	-69.70	0.7519	573.8	.000643
50000	0.1522	0.3902	0.1145	-69.70	0.7519	573.8	.000818

*THE TROPOPAUSE

CONTINUITY EQUATION

Consider the flow of air through a pipe of varying cross section as shown in figure 2.1. There is no flow through the sides of the pipe; air flows only through the ends. As the air flows through the pipe, the mass of air entering the pipe, in a given unit of time, equals the mass of air leaving the pipe, in the same unit of time. The mass flow throughout the pipe must remain constant.

The mass flow at each station is equal. Constant mass flow is called *steady-state* flow. The mass airflow is equal to the volume of air multiplied by the density of the air. The volume of air, at any station, is equal to the velocity of the air multiplied by the cross sectional area of the station.

The mass airflow, then, is the product of the density, the area, and the velocity.

$$\text{Mass airflow} = \rho AV \tag{2.6}$$

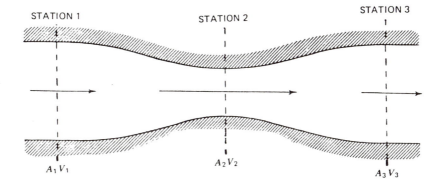

Figure 2.1 Flow of air through a pipe.

The continuity equation states that the mass airflow is constant.

$$\rho_1 A_1 V_1 = \rho_2 A_2 V_2 = \rho_3 A_3 V_3 = \text{constant} \tag{2.7}$$

The continuity equation is valid for steady state flow, both in subsonic and supersonic flow. For subsonic flow, however, the air is considered to be incompressible, and its density remains constant. The density symbols can then be canceled out and thus, for subsonic flow:

$$A_1 V_1 = A_2 V_2 = A_3 V_3 = \text{constant} \tag{2.8}$$

Velocity is inversely proportional to cross sectional area.

BERNOULLI'S EQUATION

The continuity equation explains the relationship between velocity and cross sectional area, but it does not explain differences in static pressure of the air passing through a pipe of varying cross sections. Bernoulli, using the principle of conservation of energy, developed a concept that explains the behavior of pressures in gases. Consider the flow of air through a venturi tube, as shown in Figure 2.2.

The energy of an airstream is in two forms. It has *potential energy*, which is its *static pressure*. It has *kinetic energy*, which is its *dynamic pressure*. The total pressure in the airstream is the sum of the static pressure and the dynamic pressure. The total pressure remains constant, according to the law of conservation of energy, and so an increase in one form of pressure must result in a decrease in the other.

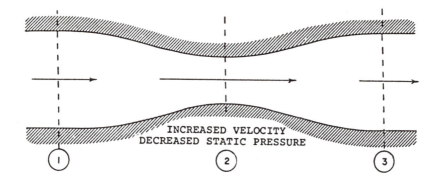

Figure 2.2   Venturi tube.

Static pressure is an easily understood concept (see the discussion earlier in this chapter). Dynamic pressure, q, is similar to kinetic energy in mechanics and is expressed by the equation:

$$q = \frac{1}{2}\rho V^2 \quad \text{(psf)}$$

(2.9)

Where V is measured in feet per second.

Pilots are much more familiar with velocity measured in knots, rather than in feet per second, so a new equation for dynamic pressure is used in this book. Its derivation is shown here:

$$\text{Density ratio, } \sigma = \frac{\rho}{\rho_o} = \frac{\rho}{0.002377}$$

$$\text{or} \quad \rho = 0.002377\sigma$$

$$V_{FPS} = 1.69(V_{KNOTS})$$

$$(V_{FPS})^2 = 2.856(V_{KNOTS})^2$$

Substituting in equation 2.9:

$$q = \frac{\sigma V^2_K}{295} \quad \text{(psf)}$$

(2.10)

Bernoulli's equation can now be expressed as:

Total pressure, H = static pressure, P + dynamic pressure, q

$$H = P + \frac{\sigma V_K^2}{295} = \text{constant}$$

(2.11)

To visualize how lift is developed on a cambered airfoil, draw a line down the middle of a venturi tube. Discard the upper half of the figure and superimpose an airfoil on the necked down section of the tube.

This is shown in Figure 2.3. Note that the static pressure is less over the airfoil than that ahead of it.

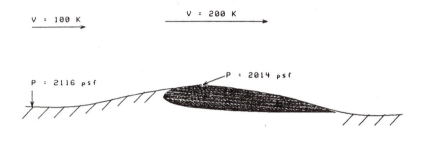

Figure 2.3   Velocities and pressures on an airfoil superimposed on a venturi tube.

AIRSPEED MEASUREMENT

If a symmetrically shaped object is placed in a moving airstream, as shown in Figure 2.4, the flow pattern will be as shown. Some of the airflow will pass over the object and some will flow beneath it, but at the point at the nose of the object, the flow will be stopped completely. This point is called the *stagnation point*. Since the air velocity at this point is zero, the dynamic pressure is also zero. The stagnation pressure is, therefore, all static pressure and must be equal to the total pressure, H, of the airstream.

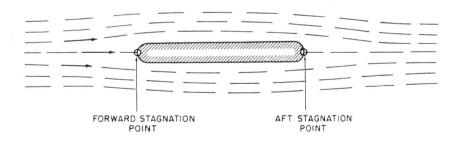

Figure 2.4   Flow around a symmetrical object.

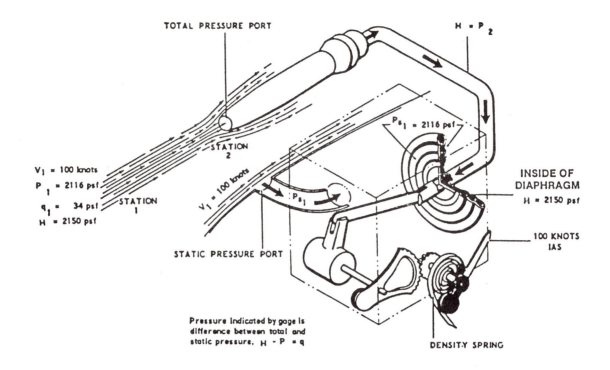

Figure 2.5   *Schematic of a Pitot-static airspeed indicator.*

In Figure 2.5 the freestream values of velocity and pressure are shown at station 1 and are for sea level standard day conditions. The pitot tube is shown as the "TOTAL PRESSURE PORT" at station 2. The pitot tube must be pointed into the relative wind for accurate readings. The air entering the pitot tube is brought to a complete stop and thus the static pressure in the tube, P_2 is equal to the total free stream pressure, H. This pressure (2150 psf) is ducted into a diaphragm.

The "STATIC PRESSURE PORT" can be made as a part of the pitot tube or, in more expensive indicators, it can be located at a distance from the pitot tube. It should be located at a point where the local air velocity is exactly equal to the airplane velocity. The static port is made so that none of the velocity enters the port, thus the port measures only static pressure, P_1, and none of the dynamic pressure, q_1. The static pressure is ducted into the chamber surrounding the diaphragm.

Now we have P_2 inside the diaphragm and P_1 outside the diaphragm. Remember that $P_2 = H = P_1 + q_1 = 2150$ psf and that $P_1 = 2116$ psf. The difference between these pressures is $H - P_1 = q_1$. This differential pressure deflects the flexible diaphragm which is geared to the airspeed pointer. The airspeed indicator is calibrated to read airspeed.

Indicated Airspeed

Indicated airspeed (IAS) is the reading of the airspeed indicator dial. If there are any errors in the instrument, they will be shown on an instrument error card, located near the instrument.

Another error, known as position error, results if the static pressure port is not located at a point on the aircraft where the local air velocity is exactly equal to the free stream velocity of the aircraft. If this error is present, another correction chart will be provided with the aircraft.

Calibrated Airspeed

Calibrated airspeed (CAS) results when the above corrections have been made to the IAS.

In fast,high altitude aircraft, the air entering the Pitot tube is subjected to a ram effect which causes the diaphragm to be deflected too far. The resulting airspeed indication is,therefore, too high and must be corrected.

Figure 2.6 shows a compressibility correction chart. A rule of thumb is that, if flying above 10,000 feet and 200 knots, the compressibility correction should be made. Unlike the instrument and position error charts, which vary with different aircraft, this chart is good for any aircraft.

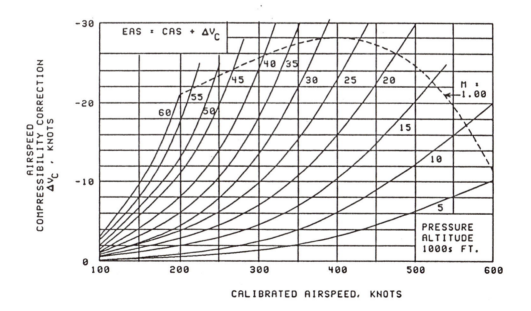

Figure 2.6   Compressibility correction chart.

Equivalent Airspeed

Equivalent airspeed (EAS) results when the CAS has been corrected for compressibility effects.

One further correction must be made to obtain true airspeed. The airspeed indicator measures dynamic pressure and is calibrated for sea level standard day density. As altitude increases, the density ratio decreases and a correction must be made. The correction factor is $\sqrt{\sigma}$.

True Airspeed

True airspeed (TAS) is obtained when EAS has been corrected for density ratio.

$$TAS = \frac{EAS}{\sqrt{\sigma}}$$

(2.12)

Values of $\sqrt{\sigma}$ can be found in the ICAO Standard Altitude Chart.

Values of $\frac{1}{\sqrt{\sigma}}$ can be found in Figure 2.7.

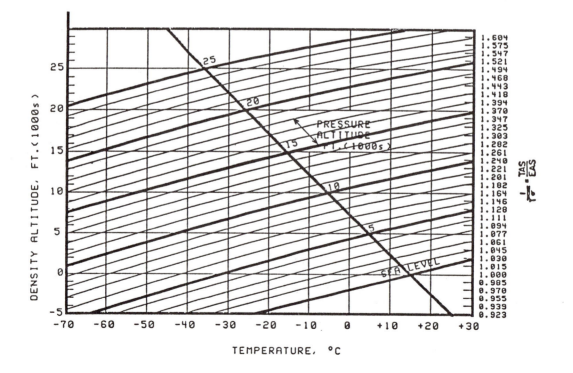

Figure 2.7   Altitude and EAS to TAS correction chart.

SYMBOLS AND UNITS

English Symbols

A	Area (ft^2)
CAS	Calibrated airspeed (knots)
°C	Celsius temperature (deg)
EAS	Equivalent airspeed
°F	Fahrenheit temperature
H	Total pressure (head) (psf)
IAS	Indicated airspeed
°K	Kelvin temperature
P	Static pressure
P_o	Sea level standard pres.
q	Dynamic pressure
R	Universal gas constant
°R	Rankine temperature
TAS	True airspeed
T	Absolute temperature
T_o	Sea level standard temp.
V	Velocity (fps)
V_K	Velocity (knots)

Greek Symbols

δ (delta)	Pressure ratio
θ (theta)	Temperature ratio
ρ (rho)	Density
σ (sigma)	Density ratio

EQUATIONS

2.1 $\quad \delta = \dfrac{P}{P_o}$

2.2 $\quad \theta = \dfrac{T}{T_o}$

2.3 $\quad \sigma = \dfrac{\rho}{\rho_o}$

2.4 $\quad \rho = \dfrac{P}{RT}$

2.5 $\quad \sigma = \dfrac{\delta}{\theta}$

2.6 $\quad$ Mass Airflow $= \rho AV$

2.7 $\quad \rho_1 A_1 V_1 = \rho_2 A_2 V_2$

2.8 $\quad A_1 V_1 = A_2 V_2$

2.9 $\quad q = \dfrac{1}{2}\rho V^2$

2.10 $\quad q = \dfrac{\sigma V^2_K}{295}$

2.11 $\quad H = P + \dfrac{\sigma V_K^2}{295}$

2.12 $\quad TAS = \dfrac{EAS}{\sqrt{\sigma}}$

PROBLEMS

1. An increase in static air pressure affects air density by:

    a. decreasing the density.
    b. does not affect the density.
    c. increasing the density.

2. A decrease in temperature affects the air density by:

    a. decreasing the density.
    b. does not affect the density.
    c. increasing the density.

3. Pressure ratio is:

    a. ambient pressure divided by sea level standard pressure measured in the same units.
    b. ambient pressure in millibars divided by 29.92.
    c. ambient pressure in pounds per square inch divided by 2116.
    d. sea level standard pressure in inches of mercury divided by 29.92.

4. Density ratio (sigma) is:

    a. equal to pressure ratio divided by temperature ratio.
    b. is measured in slugs per cubic foot.
    c. equal to the ambient density divided by sea level standard pressure.
    d. none of the above.

5. Bernoulli's equation for subsonic flow states:

    a. if the velocity of an airstream within a tube is increased, the static pressure of the air increases.
    b. if the area of a tube decreases, the static pressure of the air increases.
    c. if the velocity of an airstream within a tube increases, the static pressure of the air decreases, but the sum of the static pressure and the velocity remains constant.
    d. none of the above.

6. Dynamic pressure of an airstream is:

    a. directly proportional to the square of the velocity.
    b. directly proportional to the air density.
    c. neither a. nor b. above.
    d. both a. and b. above.

7. In this book, we use the equation for dynamic pressure, $q = \dfrac{\sigma V_K^2}{295}$, rather than the equation,

$q = \dfrac{1}{2}\rho V^2$ because:

    a. V in our equation is measured in knots.
    b. density ratio is easier to handle (mathematically) than the actual density (slugs per cubic foot).
    c. both a. and b. above.
    d. neither a. nor b. above.

8. The corrections that must be made to Indicated Airspeed (IAS) to obtain Calibrated Airspeed (CAS) are:

    a. position error and compressibility error.
    b. instrument error and position error.
    c. instrument error and density error.
    d. position error and density error.

9. The correction that must be made to CAS to obtain Equivalent Airspeed (EAS) is called compressibility error, which:

    a. is always a negative value.
    b. can be ignored at high altitude.
    c. can be ignored at high airspeed.
    d. can be either a positive or a negative value.

10. The correction from EAS to True Airspeed (TAS) is dependent upon:

    a. temperature ratio alone.
    b. density ratio alone.
    c. pressure ratio alone.
    d. none of the above.

11. An airplane is operating from an airfield which has a barometric pressure of 27.82 in. Hg. and a runway temperature of 100°F. Calculate (or find from Table 2.1) the following:

    (a) pressure ratio

    (b) pressure altitude

    (c) temperature ratio

    (d) density ratio

    (e) density altitude.

12.

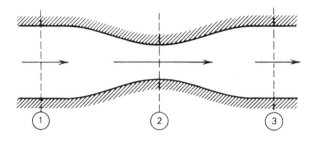

At station 1:	At station 2:	At station 3:
$A_1 = 10$ ft^2	$A_2 = 5$ ft^2	$V_3 = 80$ kts
$V_1 = 100$ kts	$V_2 = ?$	$A_3 = ?$
$P_1 = 2030$ psf	$P_2 = ?$	$P_3 = ?$
$\sigma = 0.968$		
$H = ?$	$q_2 = ?$	$q_3 = ?$

13. Using Table 2.1, calculate the dynamic pressure, q at 5000 ft density altitude and 200 knots TAS.

14. The airspeed indicator of an airplane reads 355 knots. There are no instrument or position errors. If the airplane is flying at a pressure altitude of 25,000 ft, find the equivalent airspeed (EAS).

15. Find the true airspeed (TAS) of the airplane in problem 14 if the outside air temperature is –40°C.

CHAPTER THREE

AERODYNAMIC FORCES ON AIRFOILS

AIRFOILS

An *airfoil* or, more properly, *an airfoil section*, is a slice of a wing as shown in Figure 3.1. In discussing airfoils, the planform (top view) of the wing is ignored. Wing tip effects, sweepback, taper, wash/in or wash/out, and other design features are not considered.

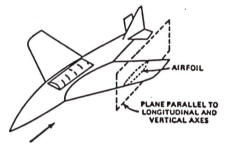

Figure 3.1   Airfoil section.

Airfoil Terminology

The terminology used to discuss an airfoil is shown in Figure 3.2.

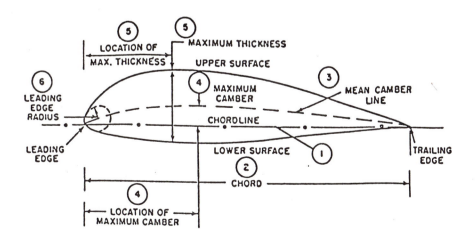

Figure 3.2   Airfoil terminology.

31

1. *Chordline* is a straight line connecting the leading edge and the trailing edge of the airfoil.

2. *Chord* is the length of the chordline. All airfoil dimensions are measured in terms of the chord.

3. *Mean camber line* is a line drawn halfway between the upper surface and the lower surface.

4. *Maximum camber* is the maximum distance between the mean camber line and the chordline. The location of maximum camber is important in determining the aerodynamic characteristics of the airfoil.

5. *Maximum thickness* is the maximum distance between the upper and lower surfaces. The location of maximum thickness is also important.

6. *Leading edge radius* is a measure of the sharpness of the leading edge. It may vary from zero, for a knife edge supersonic airfoil, to about 2 percent (of the chord) for rather blunt leading edge airfoils.

Geometry Variables of Airfoils

There are four main variables in the geometry of an airfoil:

1. Shape of the mean camber line. If the mean camber line coincides with the chordline, the airfoil is said to be symmetrical. In symmetrical airfoils, the upper and lower surfaces have the same shape and are equidistant from the chordline. A symmetrical and a cambered airfoil are shown in Figure 3.3.

2. Thickness.

3. Location of maximum thickness.

4. Leading edge radius.

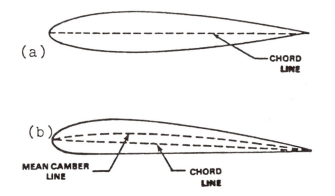

Figure 3.3   Airfoil sections: (a) Symmetrical, (b) Cambered.

DEVELOPMENT OF FORCES ON AIRFOILS

Leonardo da Vinci stated the cardinal principle of wind tunnel testing nearly 400 years before the Wright brothers achieved powered flight. Near the beginning of the sixteenth century, da Vinci stated:

"The action of the medium upon a body is the same whether the body moves in a quiescent medium, or whether the particles of the medium impinge with the same velocity upon the quiescent body."

This principle allows us to consider only relative motion of the airfoil and the air surrounding it. We may use such terms as "airfoil passing through the air" and "air passing over the airfoil" interchangeably.

DEFINITIONS

Flight Path Velocity. The speed and direction of a body passing through the air.

Relative Wind (RW). The speed and direction of the air impinging on a body passing through it. It is equal and opposite in direction to the flight path velocity.

Angle of Attack (AOA or alpha, α). The acute angle between the relative wind and the chordline of an airfoil.

Aerodynamic Force (AF). The net resulting static pressure multiplied by the planform area of an airfoil.

Lift (L). The net force developed perpendicular to the relative wind.

Drag (D). The force, parallel to the relative wind, which opposes the motion of a body through the air.

Center of Pressure (CP). The point on the chordline where the aerodynamic force acts.

Laminar Flow. Smooth airflow with little transfer of momentum between parallel layers.

Streamlined Flow. Same as laminar flow.

Turbulent Flow. Flow where the streamlines break up and there is much mixing of the layers.

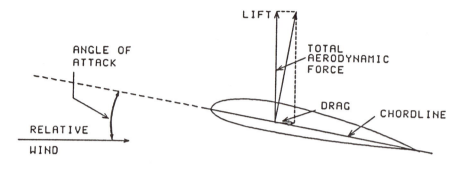

Figure 3.4   Forces acting on an airfoil.

Static Pressure Distribution on a Cylinder

Consider a stationary (non-rotating) cylinder located in a wind tunnel as shown in Figure 3.5(a). The cylinder is equipped with static pressure taps. These measure the local static pressure with respect to the ambient static pressure in the test chamber. When the tunnel is started, the airflow approaches the cylinder from the left in the figure and the static pressures are shown by the arrow vectors.

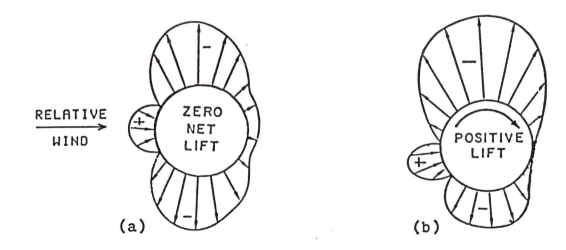

Figure 3.5   Cylinder pressures:(a) non-rotating, (b) rotating.

Arrows pointing toward the cylinder show pressures that are higher (+) than ambient and arrows pointing away from the cylinder show pressures that are less (-) than ambient static pressure.

In Figure 3.5(a) it can be seen that the upward forces are exactly resisted by the downward forces and no net vertical force (lift) is developed by the cylinder.

Now, consider that the wind tunnel is stopped and that the cylinder is rotated in the clockwise direction. Friction effects between the cylinder's surface and the air will cause some of the air to adhere to the cylinder and be rotated with it. This circular movement of the air is called *circulation*.

When the wind tunnel is restarted the air passing over the top of the cylinder will be speeded up by the circulation velocity, while the air passing over the bottom will be retarded. According to Bernoulli's equation the static pressure on the top will be reduced and the static pressure on the bottom will be increased. The new pressure distribution will be as shown in Figure 3.5(b). Lift will now be developed by the cylinder.

This force is called the *Magnus effect*. It is named after Gustav Magnus, who discovered it in 1852.

Everyday examples of the Magnus effect are seen in the curve pitch in baseball or the hook or slice of a spinning golf ball.

Pressure Disturbances on Airfoils

If an airfoil is subjected to moving airflow, velocity and pressure changes take place which create pressure disturbances in the airflow surrounding it. These disturbances originate at the airfoil surface and propagate at the speed of sound. If the flight path velocity is subsonic, the pressure disturbances, which are moving ahead of the airfoil, affect the airflow approaching the airfoil as shown in Figure 3.6.

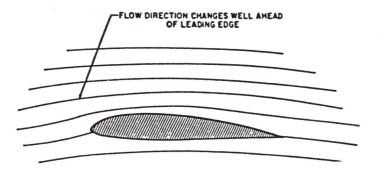

Figure 3.6   Pressure disturbances affect airflow around an airfoil.

Static Pressure Changes about an Airfoil

The air approaching the leading edge of an airfoil is first slowed down. It then speeds up again as it passes over or beneath the airfoil. At one point the air comes to a complete stop. This is the stagnation point and was discussed under the subject of airspeed measurement in Chapter Two.

Whenever the air slows down below the flight path velocity, the static pressure is higher than the ambient static pressure, in accordance with Bernoulli's equation. Likewise, whenever the local air velocity is greater than the flight path velocity, the static pressure will be less than the ambient static pressure.

At a point near maximum thickness, maximum velocity and minimum static pressure will occur. Because air has viscosity, some of its energy will be lost to friction and a "wake" of low velocity, turbulent air exists near the trailing edge. This results in a small, high pressure region in that area.

Figure 3.7(a) shows a symmetrical airfoil at zero AOA and the pressure distribution. Note that the low pressure area on the top of the airfoil is resisted by the same amount of low pressure on the bottom, hence no lift is developed.

At a positive AOA, however, as shown in 3.7(b) the airflow must speed up more as it passes over the top, resulting in a larger low pressure area. Air passing below the airfoil is slowed down and a smaller low pressure results in that region. The result is that a positive lift is developed at positive AOA.

Compare this with the pressure distribution of the non-rotating cylinder and the rotating cylinder (Figure 3.5). Even though there is no rotation of the airfoil, we call this *circulation about the airfoil.*

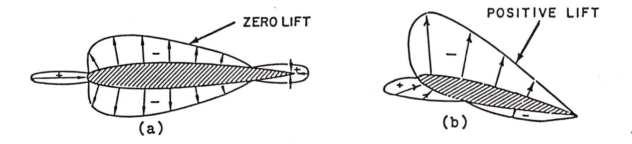

Figure 3.7   Static Pressure on an airfoil: (a) at zero AOA, (b) at a positive AOA.

AERODYNAMIC FORCE

Aerodynamic force, AF, is the resultant of all static pressures acting on an airfoil in an airflow, multiplied by the planform area that is affected by the pressure. The line of action of the AF passes through the chordline at a point called the center of pressure, CP.

It is convenient to consider that the forces acting on an airfoil, do so in some rectangular coordinate system. The system chosen is defined by the relative wind direction and an axis perpendicular to it.

The AF is resolved into two components, one parallel to the relative wind and called *Drag*, D, and the other perpendicular to the relative wind and called *Lift*, L. Figure 3.8 shows the resolution of the AF into its components L and D.

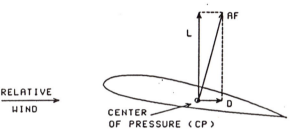

Figure 3.8   Components of aerodynamic force.

AERODYNAMIC PITCHING MOMENTS

Consider the pressure distribution about a symmetrical airfoil at zero AOA, as shown in Figure 3.9(a). The large arrows indicate the sum of the low pressures on the top and bottom of the airfoil and are located at the center of pressure of their respective surfaces.

The CP of the top of the airfoil and that on the bottom are located at the same point on the chordline, so there is no unbalance of moments about any point on the airfoil.

Figure 3.9(b) shows the pressure distribution about a symmetrical airfoil at a positive AOA. There is now an unbalance in the upper surface and lower surface lift vectors, and positive lift is being developed. However, the two lift vectors still have the same line of action, passing through the CP. There can be no moment developed about the CP. We can conclude that symmetrical airfoils do not generate pitching moments at any AOA.

While no proof is offered here, it is also true that the CP does not move with a change in AOA for a symmetrical airfoil.

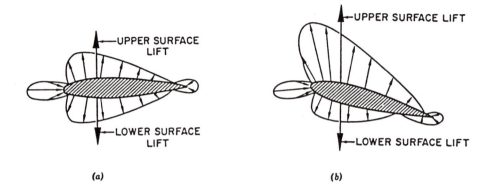

Figure 3.9 *Development of pitching moments on a symmetrical airfoil: (a) at zero AOA, (b) at positive AOA.*

Now consider a cambered airfoil operating at an AOA where it is developing no net lift, as shown in Figure 3.10(a). Upper surface lift and lower surface lift are numerically equal, but their lines of action do not coincide.

The lift on the top surface acts behind the lift on the bottom surface and a nose-down pitching moment develops. When the cambered airfoil develops lift, as shown in Figure 3.10(b), the nose-down pitching moment still exists.

By reversing the camber it is possible to create an airfoil that has a nose-up pitching moment. Delta wing aircraft have a reversed camber trailing edge to control the pitching moments.

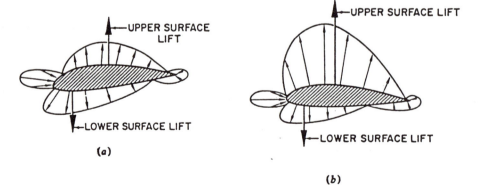

Figure 3.10 *Development of pitching moments on a cambered airfoil: (a) zero lift, (b) developing lift.*

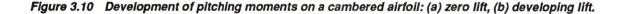

38

AERODYNAMIC CENTER

For cambered airfoils, it has been found that the CP moves along the chordline when the AOA changes. As the AOA increases, the CP moves forward and vice versa. This movement makes calculations involving stability and stress analysis very difficult.

There is a point on a cambered airfoil where the pitching moment is a constant with changing AOA, provided that the velocity is constant. This point is called the *aerodynamic center*, AC.

The AC, unlike the CP, does not move with changes in AOA, thus if we consider the lift and drag forces as acting at the AC, the calculations will be greatly simplified.

The location of the AC varies slightly, depending on airfoil shape. Subsonically, it is located between 23 and 27 percent of the chord back from the leading edge. Supersonically, the AC shifts to the 50 percent chord point.

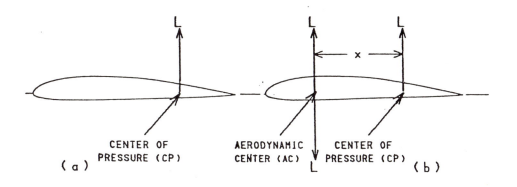

Figure 3.11   Aerodynamic Center concept.

Let's see how the lift forces can be considered to act at the AC. If we replace the upper surface lift vector and the lower surface lift vector shown in Figure 3.10(b) by a resultant "net lift" vector located at the CP, we obtain Figure 3.11(a).

Now we can place an upward vector, equal to the lift at the AC and a downward vector of the same size also at the AC. We really haven't changed anything by this maneuver. This is shown in Figure 3.11(b).

The downward vector at the AC and the upward vector at the CP create something that engineers call a *couple*. A couple does not produce any net force but does produce a moment. The moment produced by the couple is the nose-down pitching moment, M, that we discussed above. Its value is equal to the lift, L, multiplied by the distance, x, between the CP and the AC.

$$M = Lx$$

Recalling that this moment remains constant with change in AOA, it can be seen that if the AOA is reduced until the lift approaches zero, then the CP must move backward until x approaches infinity. This, of course is impossible, therefore the concept of CP is seldom used today.

In Summary:

1. The pitching moment at the AC does not change when the angle of attack changes (at constant velocity).

2. All changes in lift effectively occur at AC.

3. AC is located near 25 percent C subsonically and at 50 percent C supersonically.

SYMBOLS AND UNITS

English Symbols

AC	Aerodynamic center
AF	Aerodynamic force (lb)
AOA	Angle of attack (degrees)
CP	Center of pressure
D	Drag (lb)
L	Lift (lb)
RW	Relative wind

Greek Symbol

α (alpha)   Angle of attack (degrees)

PROBLEMS

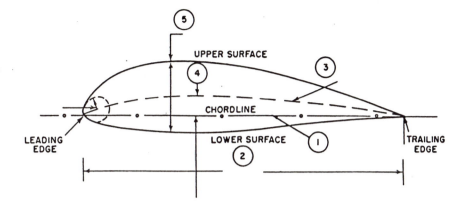

1. Number 3 on the above sketch indicates:

    a. the chord line.
    b. the maximum camber.
    c. the thickness.
    d. the mean camber line.

2. Number 4 on the above sketch indicates:

    a. the chord line.
    b. the thickness.
    c. the maximum camber.
    d. the mean camber line.

3. Number 5 on the above sketch indicates:

    a. the maximum camber.
    b. the mean camber line.
    c. the upper surface curvature.
    d. the maximum thickness.

4. The Magnus effect explains why:

    a. a bowling ball curves.
    b. a pitched baseball curves.
    c. a golf ball slices.
    d. b. and c. above.

5. Which of the below will develop positive lift?

    a. A symmetrical airfoil at zero AOA.
    b. A non-rotating cylinder in a wind tunnel.
    c. A cambered airfoil at zero AOA.
    d. None of the above.

6. Which of the below will not produce a pitching moment?

    a. A symmetrical airfoil at a positive AOA.
    b. A cambered airfoil that is developing zero lift.
    c. A cambered airfoil that is at a positive AOA.
    d. A symmetrical airfoil at zero AOA.
    e. Both a. and d. above.

7. The aerodynamic center, AC, is located at:

    a. 50% C subsonically and 25% supersonically.
    b. 25% C at all speeds.
    c. 50% C at all speeds.
    d. 25% C subsonically and 50% supersonically.

8. For a cambered airfoil, the center of pressure, CP:

    a. moves to the rear of the wing at low AOA.
    b. moves backward as AOA increases.
    c. moves forward as AOA increases.
    d. both a. and c. above.

9. Which of these statements is false? For a cambered airfoil:

    a. the AC is where all changes in lift effectively take place.
    b. there is no pitching moment at the AC.
    c. the AC is located near 25% C subsonically.
    d. the pitching moment at the AC is constant with change of AOA (at constant airspeed).

10. For a symmetrical airfoil, the center of pressure:

    a. moves forward as AOA increases.
    b. stays at the same place as AOA increases.
    c. has no pitching moment about the CP.
    d. both b. and c. above.

CHAPTER FOUR

LIFT AND STALL

AERODYNAMIC FORCE EQUATIONS

The lift and drag equations depend on several factors, the most important of which are:

1. Airstream velocity, V (knots).

2. Air density ratio, σ (dimensionless).

3. Airfoil planform area, S (ft^2).

4. Profile shape of the airfoil.

5. Viscosity of the air.

6. Compressibility effects.

7. Angle of attack, α (degrees).

The first two factors determine the dynamic pressure, q, of the airstream, as was shown in equation 2.10:

$$q = \frac{\sigma V^2}{295} \quad \text{(psf)}$$

Factors 4, 5, and 6 influence the amount of drag that an airfoil will develop at a certain AOA. It is convenient to express lift and drag forces in terms of nondimensional coefficients that are functions of AOA. These are called the coefficient of lift, C_L, and coefficient of drag, C_D.

LIFT EQUATION

$$L = C_L q S = \frac{C_L \sigma V^2 S}{295} \text{(lb)} \tag{4.1}$$

Rewriting,

$$C_L = \frac{L/S}{q} = \frac{295L}{\sigma V^2 S}$$

C_L can be said to be "the ratio between the lift pressure and the dynamic pressure." It is a measure of the effectiveness of the airfoil to produce lift. Values of C_L have been obtained experimentally for all airfoil sections and can be found in publications (see Reference 1).

Figure 4.1 shows a typical curve of C_L plotted against for a symmetrical airfoil.

Note that the curve passes through the 0-0 point, indicating that no lift is developed at 0° AOA. The curve is essentially a straight line until it approaches its stall AOA.

At the stall AOA the airfoil achieves its maximum lifting ability ($C_{L(MAX)}$). Stall is discussed later.

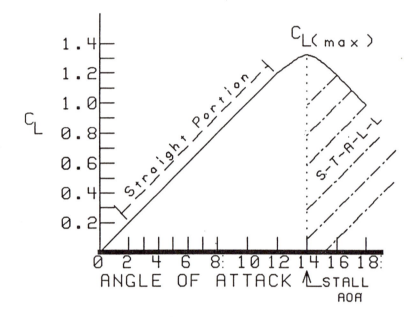

Figure 4.1 *C_L–α for a symmetrical airfoil.*

It is important to understand that an airfoil always stalls at the same AOA. For the airfoil shown in Figure 4.1 this will be at 14° AOA. The stall AOA is independent of weight, attitude and, if viscosity effects are ignored, altitude.

While the stall AOA is constant, the stall speed is not. Stall speed is affected by weight, G loading, altitude, and other factors. Rewriting the lift equation 4.1 to solve for V results in:

$$V = \sqrt{\frac{295L}{C_L \sigma S}}$$

Stall speed, V_S, occurs at $C_{L(MAX)}$, so

$$V_S = \sqrt{\frac{295L}{C_{L(MAX)}\sigma S}}$$

If the aircraft is in equilibrium, 1 G flight, the lift is equal to the weight. During maneuvering flight, however, the G loading will not be equal to the weight, but will be the weight multiplied by the Gs:

$$L = GW$$

Example: An airplane has the $C_L - \alpha$ curve of Figure 4.1 and the following data:

Gross weight = 15,000 lb.
Wing area, S = 340 ft
$C_{L(MAX)}$ = 1.3 (from Figure 4.1)

Sea level standard day.

Find the stall speed.

$$V_S = \sqrt{\frac{(295)\,(15,000)}{(1.3)\,(1.0)\,(340)}} = 100 \text{ knots}$$

Observe that, all other factors being equal, the stall speed varies as the square root of the lift (or weight under 1 G flight), so the stall speed-weight relationship is:

$$V_2 = V_1 \sqrt{\frac{W_2}{W_1}} \tag{4.2}$$

If the aircraft is flown at the same AOA, it will have the same lift coefficient, and not only the stall speed varies with the square root of the weight, but all other speeds will vary by the same amount. **Stall AOA does not vary with weight: stall speed does.**

Example: Assume the aircraft discussed above has an approach speed of 114 knots at a gross weight of 15,000 lb. It develops a lift coefficient of:

$$C_L = \frac{295W}{\sigma V^2 S} = \frac{(295)\,(15,000)}{(1.0)\,(114)^2\,(340)} = 1.0$$

Note from Figure 4.1 that a 10°AOA must be held for this flight condition. If the weight of the aircraft is increased to 20,000 lb, the AOA remains the same, but the approach speed of the aircraft must be increased, as indicated by equation 4.2:

$$V_2 = V_1 \sqrt{\frac{W_2}{W_1}} = 114 \sqrt{\frac{20,000}{15,000}} = 131.6 \text{ knots}$$

The lift equation shows that for steady unaccelerated flight the lift is equal to the weight and is a constant. There are only two variables in the equation: C_L and V_2. The value of C_L is determined by the AOA and, once this is established, the air speed is fixed. It follows then, that the control stick determines AOA and thus it also controls the airspeed. This is a fundamental concept in flying technique:

AOA is the primary control of airspeed in steady flight. Rate of climb or descent must, therefore, be controlled by throttle position.

ANGLE OF ATTACK INDICATOR

The importance of AOA in determining aircraft performance cannot be overemphasized. We have discussed stall AOA, but these facts are of equal importance: an aircraft has its maximum climb angle at a certain AOA, will achieve maximum rate of climb at another AOA, will get maximum range at still another AOA.

All aircraft performance depends on AOA, so modern aircraft are fitted with AOA indicators. This device indicates the direction of the relative wind. It consists of a probe, usually mounted on the side of the fuselage, which aligns itself with the relative wind, much like a weathervane, and a cockpit indicator. The basic cockpit indicator has a pointer showing the AOA in uncalibrated units, rather than in degrees. This allows the same instrument to be used in different aircraft.

If an AOA indicator is installed in the aircraft, the flight handbook will give the approach AOA, for instance, in units on the indicator, rather than in degrees.

Some indicators merely show approach, stall and cruise AOA's as colored segments on the indicator dial. The AOA indicating system often includes additional stall warning devices, such as stick or rudder shakers, warning horns etc.

AIRFOIL LIFT CHARACTERISTICS

The lift characteristics of airfoil sections are affected by thickness and location of maximum thickness, camber and its location, leading edge radius, and other factors. Increasing thickness, for example, results in higher local velocity airflow, producing lower static pressure and more lift. This is shown in Figure 4.2.

Both of the airfoils in Figure 4.2 have a maximum camber of 4%C, located at the 40%C position. The 4412 airfoil is 12%C thick, and the 4406 is 6%C thick. The thicker airfoil has a much higher value of $C_{L(MAX)}$, and the stall AOA is also higher.

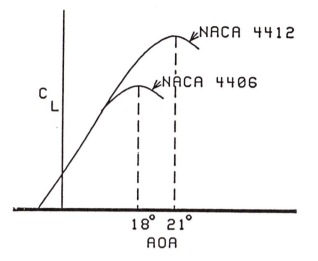

Figure 4.2   Thickness effect on C_L-α curves.

The 4412 airfoil is superior to the 4406 airfoil in producing lift, but it will also have more drag. The mission of a proposed aircraft will influence the type of airfoil selected. Cambered airfoils produce lift at zero AOA, as can be seen from the above figure.

Figure 4.1 shows the lift coefficient curve for a symmetrical airfoil and Figure 4.2 shows the curves for two cambered airfoils. To better compare the two types, we can show them together, as in Figure 4.3. An increase in $C_{L(MAX)}$ and a decrease in stall AOA will be noted when camber is increased.

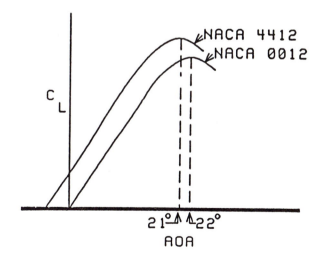

Figure 4.3   Camber effect on C_L-α curves.

BOUNDARY LAYER THEORY

As we have seen, high AOA produces high coefficients of lift. Thus lift to support an aircraft can be obtained at low airspeeds. To better understand what happens to the airflow when the aircraft is flown at or above stall AOA, we must understand the behavior of the air close to the airfoil (or other aircraft) surface.

Consider an airfoil in a wind tunnel. The viscosity of the air causes the air particles next to the surface to adhere to the surface, thus their velocity is zero. Particles of air slightly farther away from the surface will be slowed down, because of friction effects between the particles, but will not be completely stopped. The farther we move away from the surface, the faster the particles will be moving until at some distance from the surface a point is reached where the particles are not slowed up at all. The layer of air from the surface of the airfoil to the point where there is no measurable slowing up of the air, is known as the *boundary layer*. The nature of the boundary layer determines the *maximum* lift coefficient and the stalling characteristics of the airfoil.

The beginning of airflow at the leading edge of a smooth airfoil surface produces a very thin layer of smooth airflow. This type of airflow is called *laminar flow* and is characterized by smooth regular streamlines. Fluid particles in this region do not intermingle.

As the airflow moves back from the leading edge, the boundary layer thickens and becomes unstable. Small pressure disturbances cause the unstable airflow to tumble, and intermixing of the air particles takes place. This type of airflow is called *turbulent flow.*

Figure 4.4 illustrates the flow within the boundary layer on a flat plate and shows the increasing thickness of the boundary layer and transition from laminar to turbulent flow. The thickness is greatly exaggerated in the drawing.

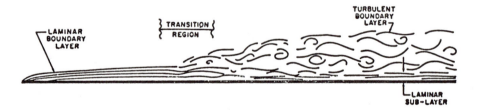

Figure 4.4 Boundary layer composition.

You can observe laminar and turbulent airflow by lighting a cigarette in a draft-free room and noting the smoke as it rises, as shown in Figure 4.5.

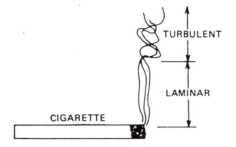

Figure 4.5   Smoke pattern.

Velocity profiles are drawn to help visualize the local velocity of an airstream in the boundary layer.

Figure 4.6 shows one of these. The relative velocity at the surface is zero, and the relative velocities at various levels away from the surface increase until the outer edge of the boundary layer is reached.

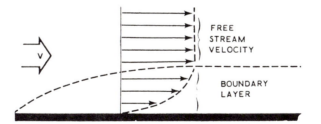

Figure 4.6   Velocity profile in boundary layer.

Velocity profiles are different in laminar and turbulent flow.

Laminar profiles show a gradual increase in the relative velocity from the surface to the outer edge of the boundary layer.

Turbulent flow involves rapid intermixing of the air levels, and faster moving air speeds up the particles near the surface, resulting in profiles as shown in Figure 4.7.

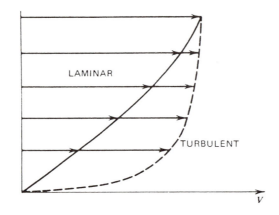

Figure 4.7   Laminar and turbulent velocity profiles.

REYNOLDS NUMBER

In 1883 Osborn Reynolds carried out a series of experiments with the flow of water in pipes. By injecting dye into a stream of flowing water, he discovered that laminar and turbulent flow could be seen. He also discovered a relationship between the point where the fluid became turbulent and three properties of the fluid stream.

Reynolds number is defined by the following equation:

$$R_E = \frac{V_x}{v}$$

(4.3)

where

R_E = Reynolds number
V = stream velocity (fps)
x = distance downstream (ft)
v = kinematic viscosity (ft^2 /sec)

Reynold's experiments showed that at low values of R_E the flow was laminar, and at higher values the flow was turbulent. When considering airflow about an airfoil, it is not possible to define the exact R_E when flow changes from laminar to turbulent. The contour of the airfoil and the smoothness of the surface influence the changeover point. Studies on smooth flat plates have been made, and a prediction of the type of flow can be made. Figure 4.8 shows the results of such a study: laminar airflow exists below R_E of 0.5 million, and turbulent flow exists for values greater than 10 million. Transition occurs between these two values.

Reynolds numbers corrections are very important in correlating wind tunnel data of scale models with full-sized aircraft.

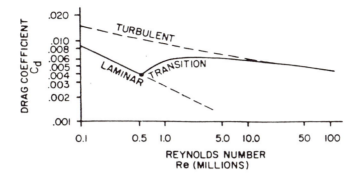

Figure 4.8   Reynolds number effect on airflow on a smooth flat plate.

ADVERSE PRESSURE GRADIENT

The pressure distribution about a cambered airfoil which is producing lift is shown in Figure 4.9.

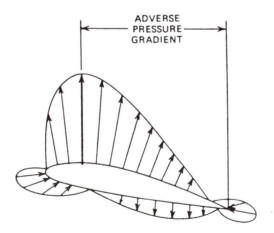

Figure 4.9   Adverse pressure gradient.

The minimum pressure point is shown by the heavy arrow. Airflow approaching the leading edge of the airfoil enters the high pressure area near the stagnation point. As the air divides to pass over the top and under the bottom surfaces of the airfoil, it moves from a high pressure area to a lower static pressure area. The air moving over the top of the airfoil is of most importance to this discussion. Until the air reaches the point of minimum pressure, it is in a favorable pressure gradient. The forces of nature are acting on the air in a favorable manner, and it is being accelerated and sped on its way.

Once the air on the top of the airfoil passes the minimum static pressure point, however, it is entering a region of higher static pressure. This is contrary to the laws of nature, and this region is called an *adverse pressure gradient*. The kinetic energy of the airflow is now being transformed into higher static pressure, and the velocity is being reduced.

AIRFLOW SEPARATION

The airflow in the boundary layer is being acted on by two forces:
   (1) friction forces and
   (2) adverse pressure gradient forces.

First, the friction forces between the surface of the airfoil and the air particles, and the friction between the particles themselves, both tend to reduce the relative velocity to zero. Second, the air is being slowed by the adverse pressure gradient.

The result of the air slowing creates a stagnated region close to the surface. The air in this region is said to be aerodynamically "dead."

Airflow from outside the boundary layer will overrun the point of stagnation and a flow reversal results. This means that the airflow moves forward, the boundary layer separates from the surface, lift is destroyed, and drag becomes excessively high.

Airflow can be seen on an airfoil operating at high AOA if tufts of wool are glued to the upper surface. The tufts will reverse direction and flap violently when stall occurs.

Figure 4.10 shows the velocity profiles in the boundary layer during airflow separation.

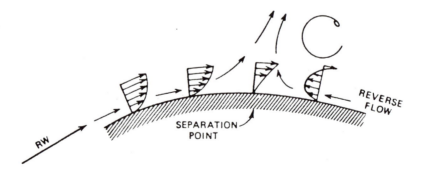

Figure 4.10   Airflow separation velocity profiles.

Consider a smooth sphere in an airflow as shown in Figure 4.11(a). If the diameter is small, the Reynolds number will be small, and laminar airflow will exist on the surface. This produces only a small amount of drag against the surface, but laminar flow allows the air to separate easily, and a wide, very high drag wake will form behind the sphere.

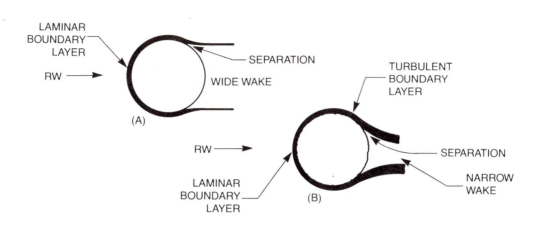

Figure 4.11   Sphere wake drag: (a) smooth sphere, (b) rough sphere.

Now, assume the sphere's surface is roughened, as shown in Figure 4.11(b). The airflow becomes turbulent very soon after it passes over the sphere. The surface drag is now increased, but turbulent air has more energy than laminar flow and will resist separation. Thus the width of the wake is greatly reduced, and overall drag is reduced.

The dimples on a golf ball and the fuzz on a tennis ball are examples of how creating turbulent flow delays separation and reduces total drag.

STALL

Stall is airflow separation of the boundary layer from a lifting surface. It is characterized by a loss of lift and an increase in drag. There are two types of stall that are of interest to the non-jet pilot.

The most common stall is the slow speed, high AOA separation. It most often occurs during the takeoff and landing phases of flight and is dangerous because of the lack of altitude for recovery.

This type of stall starts at the trailing edge of a wing and progresses forward. The part of the wing that stalls first is of great interest and depends on the planform shape of the wing. This will be discussed in detail later.

The second kind of stall is the *accelerated stall*. It occurs when a sudden rapid increase in AOA is made. Sharp leading edges contribute to this type of stall. The rapid increase in AOA does not allow time for the airflow to turn the corner of the leading edge, and separation takes place there. To avoid this type of stall do not make sudden nose-up movements of the control stick in recovering from dives or unusual attitudes.

HIGH COEFFICIENT OF LIFT DEVICES

Landing and takeoff flaps, slots, slats, and other devices that enable an aircraft to take off and land at reduced speeds are often erroneously called "high lift devices." The lift of the aircraft during these maneuvers does not exceed its weight; therefore, high lift is not required. However, landing and takeoff speeds can be reduced by increasing the coefficient of lift, so the proper term is "high coefficient of lift" devices.

In our discussion of airfoil lift characteristics we saw two ways that $C_{L(MAX)}$ of an airfoil could be increased in value:
    (1) by increasing its thickness and
    (2) by increasing its camber.

No practical way of increasing its thickness in flight has been developed to date, so this possibility is not discussed here. The first attempts to provide high C_L devices concentrated on increasing the camber of the wing. Devices which increase C_L by this method are classified as camber changers.

Some examples of these are shown in Figure 4.12.

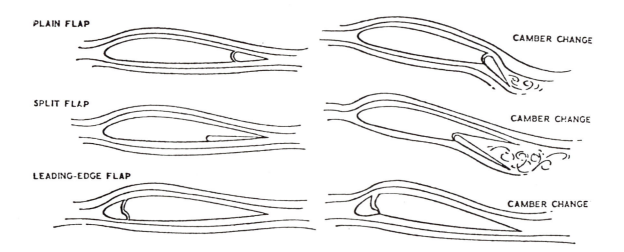

Figure 4.12   Pure camber changers.

Figure 4.13 shows the effects of pure camber changers on the coefficient of lift curves.

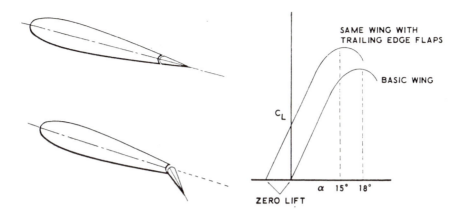

Figure 4.13   Effect of a "camber changer" on C_L-α curve.

Our discussion of boundary layer theory, separation, and stalls revealed that airflow separation resulted when the lower levels of the boundary layer did not have sufficient kinetic energy and were slowed by friction and an adverse pressure gradient.

If we can find a way to increase the kinetic energy in the boundary layer, flow separation can be suppressed, higher angles of attack can be used, and higher values of C_L can be realized. Devices that use this principle are called *energy adders.*

There are several ways of adding energy to the boundary layer. The simplest seems to be by stirring up the boundary layer, thus replacing the slow moving laminar airflow next to the wing skin with faster moving turbulent air.

One method of producing turbulence in the boundary layer is the use of vortex generators. These are small airfoil-shaped vanes, protruding upwards from the wing as shown in Figure 4.14.

Vortex generators mix the turbulent outer layers of the boundary layers with the slow moving laminar lower layers thus reenergizing them.

The advantages of reduced landing and takeoff speeds, due to the increase in lift coefficient, more than compensate for the increased parasite drag of the vortex generators.

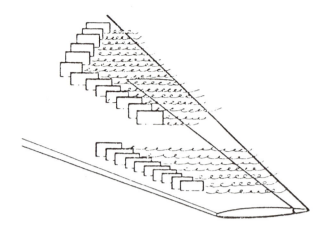

Figure 4.14   Vortex generators.

A second method of adding kinetic energy to the boundary layer is to route high pressure air from the stagnation region, near the leading edge of the wing, to the low pressure area on top of the wing, where the lower layers of the boundary layer are slowing up. Figure 4.15 shows this method, using a fixed slot at the leading edge of the wing.

At low angles of attack the pressure on each end of the slot is about the same and little or no air flows through the slot as shown in Figure 4.15(a). At high angles of attack, however, the stagnation point has moved well down from the leading edge, and a high pressure differential causes a large airflow through the slot and over the top of the wing, as shown in Figure 4.15(b).

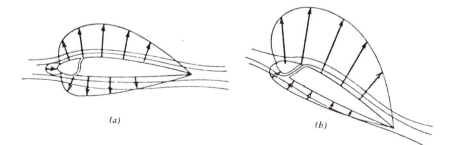

(a)   (b)

Figure 4.15   Fixed slot at:(a) low AOA, (b) high AOA.

The effect of pure energy adders on the C_L curve is shown in Figure 4.16. Unlike the camber changer devices, which move the entire curve upward (Figure 4.13), the energy adders extend the curve to higher values of AOA and $C_{L(MAX)}$ without increasing the value of C_L for lower values of AOA.

Most of the more complicated high C_L devices use a combination of camber changer and energy adder techniques.

Slotted leading and trailing edge flaps are examples of combination devices.

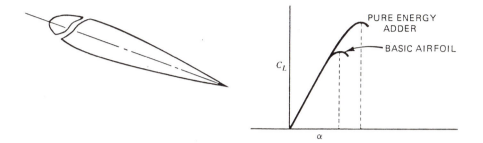

Figure 4.16   Effect of an "energy adder" on the C$_L$-α curve.

A third method of increasing the kinetic energy in the boundary layer is by supplying energy from some outside source. This type device is called boundary layer control, (BLC). There are two methods of doing this.

The first is by use of suction. A vacuum pump is attached to ducts, which lead to spanwise slots in the upper surface of the wing. When suction is applied, the lower layers of air in the boundary layer are removed and are replaced by faster moving air from the higher layers. This delays separation of the boundary layer from the top of the wing. As there is no available source of suction on the airplane, this method is seldom, if ever, used.

The most common type of BLC is the "blowing air" device. High pressure air is extracted from the compressor of the turbine engine and piped into rearward facing slots located in front of the trailing edge flaps. This blowing air re-energizes the boundary layer and delays separation. Figure 4.17 shows this type of BLC.

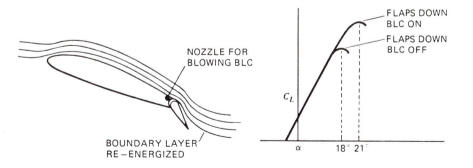

Figure 4.17   Effect of blowing BLC over a trailing edge flap on the C$_L$-α curve.

SYMBOLS AND UNITS

English Symbols

C_L	Coefficient of lift (dimensionless)
$C_{L(MAX)}$	Maximum value of C_L
G	Load factor (dimensionless)
R_E	Reynolds number (dimensionless)
S	Wing Area (ft^2)

Greek Symbol

ν (nu) Kinematic Viscosity (ft^2/sec)

EQUATIONS

4.1 $$L = \frac{C_L \sigma V^2 S}{295}$$

4.2 $$V_2 = V_1 \sqrt{\frac{W_2}{W_1}}$$

4.3 $$R_E = \frac{V_x}{\nu}$$

PROBLEMS

1. As thickness of an airfoil is increased, the stall AOA:

    a. is greater.
    b. is less
    c. remains the same.

2. As camber of an airfoil is increased, its C_L at any AOA:
    a. is less.
    b. remains the same.
    c. is greater.

3. Air in the boundary layer:

    a. has zero velocity at the wing surface.
    b. is turbulent near the leading edge.
    c. is more apt to stall if it is turbulent.
    d. none of the above.

4. Air in the boundary layer:

    a. changes from laminar to turbulent at low Reynolds numbers.
    b. separates from the wing when its velocity is maximum.
    c. reverses flow direction when stall occurs.
    d. none of the above.

5. Two things an airfoil designer can change to increase $C_{L(MAX)}$ are:

    a. thickness and wing area.
    b. chord length and aspect ratio.
    c. camber and wing span.
    d. thickness and camber.

6. High values of Reynolds number will more likely:

    a. occur near the leading edge of an airfoil.
    b. indicate laminar flow.
    c. occur at lower airspeeds.
    d. indicate turbulent airflow.

7. Adverse pressure gradient on an airfoil is found:

    a. from the point of maximum thickness to the trailing edge.
    b. near the stagnation point at the leading edge.
    c. from the point of minimum pressure to the trailing edge.
    d. both a. and c. above.

8. At high AOA stall, the air flow:

    a. stops.
    b. reverses direction.
    c. speeds up.
    d. moves toward the wing tips.

9. Airflow separation can be delayed by:

    a. making the wing surface rough.
    b. using vortex generators.
    c. directing high pressure air to the top of the wing or flap through slots.
    d. all of the above.

10. Indicate which type of high C_L device (camber changer, energy adder, or combination) each of the below is.

    a. blowing air (BLC) over a flap. _____
    b. fixed slot. _____
    c. slotted leading edge flap. _____
    d. plain flap. _____

11. An aircraft has the C_L curve shown in Figure 4.1. The following data applies:

    weight   = 20,000 lb.

    wing area, S   = 340 ft^2.

    density altitude = 10,000 ft.

    TAS   = 242.4 knots.

Find the aircraft's AOA for steady flight.

12. Calculate the stall speed for the above aircraft.

13. If the airplane in problem 11 burns up 5000 lb. of fuel during its flight, calculate its new stall speed.

14. An airplane has a rectangular wing with a chord length of 8 ft. If the airplane is flying at 200 knots TAS and at 20,000 ft. density altitude, find the Reynolds number at the trailing edge of the wing.

CHAPTER FIVE

DRAG

Drag is the component of the aerodynamic force that is parallel to the relative wind and retards the forward motion of the aircraft. At subsonic speeds, there are two kinds of drag: *induced drag* and *parasite drag*.

Shock wave drag is a third kind of drag, occurring at high speeds.

DRAG EQUATION

The basic drag equation is quite similar to the basic lift equation 4.1. It is:

$$D = C_D qS = \frac{C_D \sigma V^2 S}{295} \tag{5.1}$$

Rewriting:

$$C_D = \frac{D/S}{q} = \frac{295D}{\sigma V^2 S}$$

The coefficient of drag, C_D, can be said to be "the ratio of the drag pressure to the dynamic pressure." Values of C_D have been obtained experimentally for all airfoil sections and can be found in publications (see ref. 1). Figure 5.1 shows a typical curve for C_D-α and C_L-α for the same airfoil.

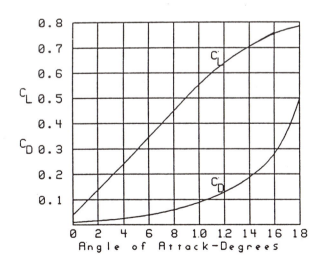

Figure 5.1 *C_L-α and C_D-α.*

LAMINAR FLOW AIRFOILS

In the late 1930's a new radical design of the airfoil was discovered. This is known as the Laminar flow airfoil.

Two symmetrical airfoils, one of the National Advisory Committee for Aeronautics (NACA) "4-digit series" and one of the NACA "6-series" (laminar flow), are compared using C_D-C_L curves in Figure 5.2.

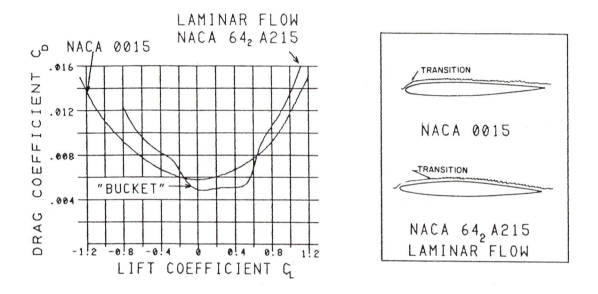

Figure 5.2 *Comparison of drag characteristics of conventional and laminar flow airfoils.*

The most obvious difference in these curves is the "drag bucket" that exists for the laminar flow airfoil. As long as this airfoil operates at low C_L (low AOA), the airflow remains laminar for a much farther distance back from the leading edge than on the conventional airfoil.

As we have seen, laminar flow has much less skin friction drag than turbulent flow, and thus less C_D.

At higher C_L values, the advantage of the laminar airfoil disappears. The sharper leading edge of the laminar airfoil causes early separation and a resulting increase in drag coefficient.

This airfoil was used on the B-24 Liberator bombers in World War II. It was not an overwhelming success for several reasons.

First, flying from muddy dirt airfields resulted in mud being thrown on the leading edges of the wings. This disrupted the laminar flow.

Shrapnel damage had a similar effect.

Finally, loss of one or two engines required that the airplane fly at low airspeeds and high angles of attack; thus flight in the "drag bucket" was impossible.

LIFT/DRAG RATIO

An aircraft's lift/drag ratio (L/D) is a measure of its efficiency. An aircraft with a high L/D is more efficient than one with a lower L/D.

The L/D can be found by dividing the values of C_L by those of C_D, at the same AOA. This can be seen if we divide equation 4.1 by 5.1. All the factors cancel out except L, D, and their coefficients, so

$$L/D = C_L/C_D \qquad (5.2)$$

Typical L/D curves are plotted in Figure 5.3.

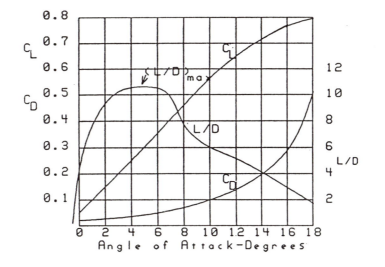

Figure 5.3   Typical lift to drag ratios.

The highest point on the curve is of great importance. It is called $(L/D)_{MAX}$ and occurs at the most efficient AOA. Minimum drag occurs at $(L/D)_{MAX}$. Under 1 G flight, the lift of the aircraft equals the weight, so L/D = W/D. If the value of W/D is a maximum, then drag must be a minimum. Minimum drag is then equal to the weight of the aircraft divided by the value of $(L/D)_{MAX}$.

$$D_{MIN} = \frac{W}{(L/D)_{MAX}} \qquad (5.3)$$

Several performance items occur at $(L/D)_{MAX}$. One of these is that the best engine(s) out glide ratio is achieved here. As lift is proportional to distance and drag is proportional to altitude, then L/D is proportional to glide ratio (distance/altitude). This means that the numerical value of L/D = glide ratio.

Note that the value of $(L/D)_{MAX}$ and the angle of attack at which it occurs do not vary with aircraft weight or altitude. The airspeed for $(L/D)_{MAX}$ does vary with weight, however, as was seen in equation 4.2.

INDUCED DRAG

Induced drag, D_i, is the least understood type of drag, but it is the most important, especially in the critical low speed region of flight. It is often called "drag due to lift" because it only occurs when lift is being developed.

Up to this point we have been discussing airfoils rather than wings. We must now consider the shape of the wing as viewed from above. This is called the *planform* of the wing. In describing the planform of a wing, several terms must be explained. These are illustrated in Figure 5.4.

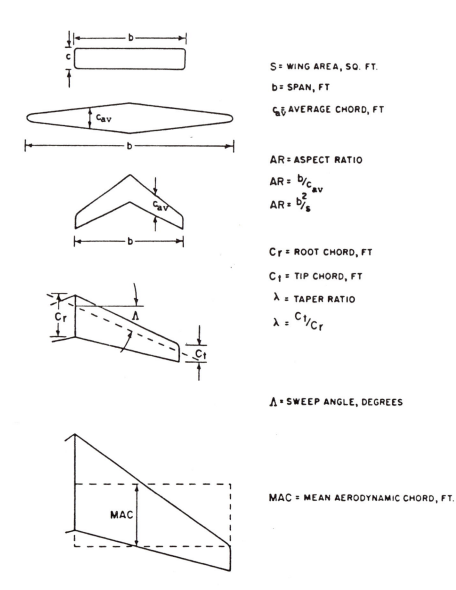

S = WING AREA, SQ. FT.

b = SPAN, FT

c_{av} = AVERAGE CHORD, FT

AR = ASPECT RATIO

$AR = b/c_{av}$

$AR = b^2/s$

C_r = ROOT CHORD, FT

C_t = TIP CHORD, FT

λ = TAPER RATIO

$\lambda = C_t/C_r$

Λ = SWEEP ANGLE, DEGREES

MAC = MEAN AERODYNAMIC CHORD, FT.

Figure 5.4   Wing planform terminology.

1. The *wing area*, S, is the plan surface area including the area covered by the fuselage or nacelles.

2. The *wing span*, b, is the distance from tip to tip.

3. The *average chord*, C_{AV}, is the geometric average chord.
   Span x average chord = wing area (bC_{AV} = S).

4. The *aspect ratio*, AR, is the span divided by the average chord. $AR = \dfrac{b}{C_{AV}}$; also $AR = \dfrac{b^2}{S}$.

5. The *root chord*, C_R, is the chord measured at the aircraft centerline. The tip chord, C_T, is the chord measured at the wing tip.

6. The *taper ratio*, λ (lambda), is the ratio of the tip chord to the root chord. $\lambda = \dfrac{C_T}{C_R}$.

7. The *sweep angle*, Λ (LAMBDA), is measured as the angle between the line of 25 percent chord points and a perpendicular to the root chord.

8. The *mean aerodynamic chord*, MAC, is the chord drawn through the centroid (center of area) of the halfspan area. The MAC and the C_{AV} are **NOT** the same. If the actual wing could be replaced by a rectangular wing having the same span and a chord equal to the MAC, the pitching moments of each would be identical.

Wing Tip Vortices

Wing tip vortices are formed when higher pressure air below a wing flows around the wing tips into the lower pressure region on top of a wing that is developing lift. The vortices are strongest at the tips and become weaker progressing toward the centerline of the aircraft, as shown in Figure 5.5.

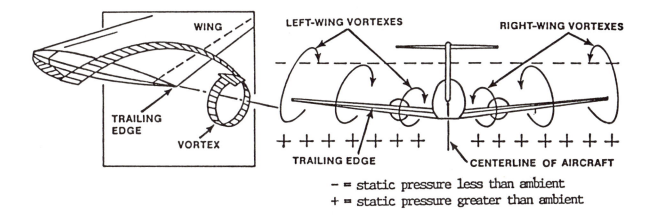

$-$ = static pressure less than ambient
$+$ = static pressure greater than ambient

Figure 5.5   Wing tip vortices.

Drag

Consider an aircraft with an infinitely long wing. This *infinite* wing has no wing tips and, consequently, no wing tip vortices. The absence of wing tip vortices means that the upwash in front of the wing and the downwash behind the wing cancel each other, as shown in Figure 5.6, and there is no net downwash behind the wing.

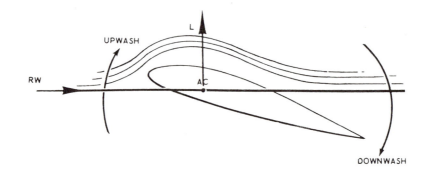

Figure 5.6   Airflow about an infinite wing.

The relative wind RW ahead of this wing is horizontal, and the RW behind the wing is also horizontal. The RW at the aerodynamic center AC is the average of these two and, therefore, must also be horizontal. Lift is 90° to the RW, so it acts vertically.

The vertical components of the local RW for an infinite wing are plotted in Figure 5.7. Note that the RW at the AC has no upward or downward velocity and, therefore, is horizontal.

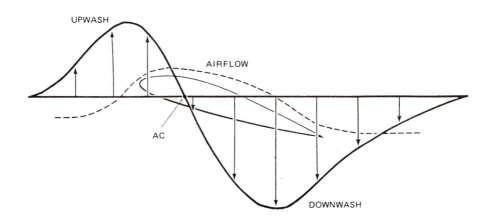

Figure 5.7   Vertical velocity vectors of infinite wing.

A wing with wing tips is called a *finite* wing. A finite wing that is developing lift will have tip vortices, as described previously. These vortices force air down behind the wing. The vertical components of the local RW for a finite wing that is producing lift are shown in Figure 5.8.

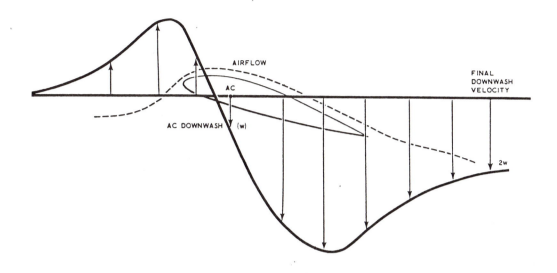

Figure 5.8   Vertical velocity vectors of finite wing.

The local RW is the vector sum of the free stream RW and the vertical velocity vectors, as shown in Figure 5.8. Far ahead of the aircraft, the RW is not disturbed by upwash and so, for an aircraft flying horizontally, it is also horizontal. Behind the wing, the local RW is depressed by the final downwash velocity (2w). The angle that the RW leaving the wing makes with the free stream RW is called the *downwash angle* and can be seen in Figure 5.9

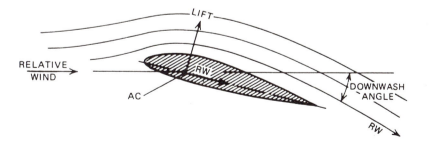

Figure 5.9   Airflow about a finite wing.

The local RW at the AC is the average of the approaching and departing relative winds. The downwash angle of the local RW at the AC is, therefore, one-half of the trailing edge downwash angle. The lift vector at the AC, which is perpendicular to the local RW, is tilted back by this same angle. The angle that the lift vector is tilted backward is called the *induced angle of attack*, α_i.

The finite wing is operating at a lower AOA than the infinite wing due to the depressed RW at the AC. Thus, if the finite wing is to produce the same lift coefficient as the infinite wing, the AOA, with respect to the free stream RW, must be increased by α_i. This is shown vectorially in Figure 5.10. The AOA of the infinite wing is α_o; the AOA of the finite wing is α.

$$\alpha = \alpha_o + \alpha_i \tag{5.4}$$

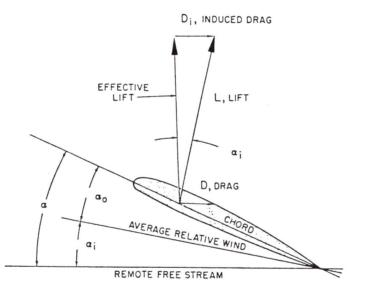

Figure 5.10 *Relative wind and force vectors on a finite wing*

The vertical vector, Effective Lift, is the lift vector of the infinite wing, and the tilted vector, L, Lift, is the lift vector of the finite wing. The component of the tilted lift vector that is parallel to the free stream RW is the induced drag, D_i.

Induced drag is influenced by lift coefficient and aspect ratio. It increases directly as the square of C_L and inversely as the aspect ratio. Thus, the most critical conditions, where induced drag is greatest, is during low speed (high C_L) flight with a low aspect ratio (short wingspan) aircraft.

Induced drag varies inversely with the velocity squared as shown in Figure 5.11.

$$\frac{D_{i_2}}{D_{i_1}} = \left(\frac{V_1}{V_2}\right)^2$$

(5.5)

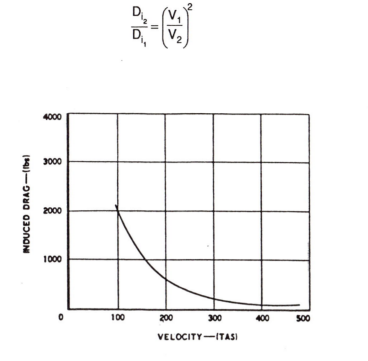

Figure 5.11   Induced drag versus velocity.

Induced Drag Summary

1. Wing tip vortices are formed by higher pressure air beneath a wing moving into the lower pressure air above the wing.

2. Wing tip vortices cause the airflow behind the wing to be pushed downward; this is called downwash.

3. Downwash causes the RW behind the wing to be deflected downward at a downwash angle.

4. The RW at the AC is influenced by the downwash, and it is deflected downward by one-half of the downwash angle.

5. The lift vector is tilted backward by the induced angle of attack, α_i, which is numerically equal to one-half of the downwash angle.

6. The rearward component of the tilted lift vector is induced drag.

GROUND EFFECT

There are three possible effects that occur when an airplane is flown close to the runway or to a water surface.

These are collectively called *ground effect* and are noticeable when the aircraft is about one wing span or less above the surface. The closer the aircraft is to the surface, the greater the effects will be.

The most noticeable effect is caused by the surface as it physically reduces the downwash behind the wing.

This, in turn, reduces both the upwash and the wing tip vortices. Without so much downwash, the local relative wind at the aerodynamic center is more nearly horizontal. The lift vector is more nearly vertical, and the induced drag is greatly reduced. The reduction in induced drag coefficient as a function of height above the runway is shown in the graph in the center of Figure 5.12.

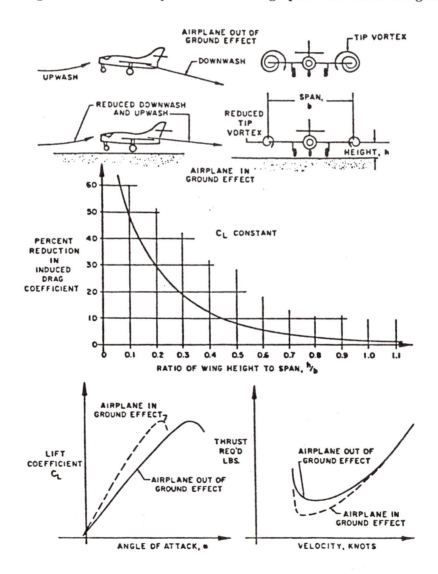

Figure 5.12   Ground effect.

70

Low winged aircraft are more affected by ground effect than high winged aircraft. This reduction of drag is clearly seen on light aircraft in their tendency to "float" during landings. To reduce this floating tendency, the aircraft approach should be made with some throttle on, rather than at idle, then close the throttle as the aircraft is flared. Thus the reduction of drag is compensated for by a reduction in thrust, and there is little or no floating action.

Ground effect changes the C_L-α curves and the thrust required curves as can be seen in Figure 5.12. The reduction of thrust required means a reduction in fuel flow. This effect has been used in emergency flying over oceans, when the fuel available was critical.

A word of caution to pilots of turbine engine aircraft: Fuel consumption is high for these aircraft at low altitude, so the benefits of ground effect may be nullified by poor fuel consumption.

Other less known effects from flying in ground effect influence many aircraft. Nose-up pitching moments are required to rotate an aircraft to an increased AOA for takeoff or to flare the aircraft during landing. This requires a down load on the tail. To obtain this pitching moment the pilot applies back stick, and the horizontal tail is then effectively operating at a negative AOA. The down-wash from the wing accentuates this negative AOA.

When the aircraft enters ground effect, on landing, the downwash is decreased, the negative AOA is reduced, the downward lift on the tail is reduced, and the aircraft experiences a nosedown pitching moment. High tailed aircraft will not experience this pitching moment, if the tail is above the wing downwash and is not influenced by it.

During takeoff, as the aircraft leaves ground effect, the opposite pitching moment occurs. Here, the increased downwash hits the horizontal tail and causes a down load on it and a nose-up pitching moment results. Unless corrective action is taken, this will cause a sharp increase in drag, which when coupled with the increase in induced drag that normally results when an aircraft leaves ground effect, may prevent a successful takeoff.

A third result from flying in ground effect is experienced by aircraft that have static airspeed ports located below the wings. There is a small but measurable increase in static pressure below the wings as an aircraft enters ground effect. This creates a false reading from the airspeed indicator. It will read lower than the actual airspeed on entering ground effect. Normally this goes unnoticed during takeoffs and landings, but on long flights over water, when trying to take advantage of ground effect, it may upset a pilot to see the airspeed reduce as the aircraft approaches the water.

On entering ground effect:
    (1) Induced drag is decreased,
    (2) nose-down pitching moments occur, and
    (3) the airspeed indicator reads low.

On leaving ground effect:
    (1) Induced drag is increased,
    (2) nose-up pitching moments occur, and
    (3) the airspeed indicator will read higher (correctly).

PARASITE DRAG

Parasite drag, D_p, is easily understood, but difficult to measure. This is because there are several types of parasite drag and the total parasite drag is not a simple addition of each component. Five types of parasite drag are discussed here.

1. *Skin friction drag* is caused by the viscous friction within the boundary layer. The total area of the aircraft skin that is exposed to the airstream will be affected by this type of drag. Skin smoothness also greatly affects this drag. Flush head rivets, waxed and polished surfaces, and removal of aluminum oxide help reduce skin friction drag.

2. *Form drag* is the portion of the parasite drag that is influenced by the shape or form of the aircraft. Streamlining the fuselage, engine nacelles, pods, and external stores helps reduce form drag.

3. *Interference drag* is caused by the interference of boundary layers from different parts of the airplane. If the drag of two component parts of an airplane are measured individually and then the parts are assembled, the drag of the assembly will be greater than the drag of the parts. The boundary layer interference is the reason for this. Smooth fairings at surface junctions reduce this type of drag.

4. *Leakage drag* is caused by differential pressure inside and outside of the aircraft. Air flowing from a higher pressure inside the fuselage through a crack or door seal will create an airstream which impinges on the airflow around the aircraft and creates drag. Door and window sills are often sealed with masking tape before starting an air race to lessen this drag.

5. One other kind of drag, of particular interest to helicopter pilots, is called *profile drag*. Profile drag is the drag of the moving rotors, and it develops anytime the rotors are in motion. So this drag can exist even if the aircraft is not in motion or developing lift.

Parasite Drag Calculations

As was mentioned above, parasite drag is hard to measure, but there is a way of simplifying this measurement.

Consider the basic drag equation 5.1 with subscripts, $_p$, added to indicate "parasite."

$$D_p = C_{Dp}qS$$

At any AOA the value of C_{Dp} will be a constant and, of course, S is a constant. A new constant, called the equivalent parasite area, f, is now introduced.

$$f = (C_{Dp})(S)$$

Because C_{Dp} is dimensionless, the dimensions of f are the same as S, (ft^2). The equivalent parasite area is often called the "barn door area" of the aircraft. This is not technically correct, but it may help you visualize the area as that seen by an observer of an approaching aircraft.

The value of f for an aircraft can be determined by placing a model in a wind tunnel and adjusting the AOA until zero lift is being developed, then measuring the drag. With no lift, there is no induced drag, and so all the drag will be parasite. If the density ratio, and tunnel velocity are known, then the equivalent parasite area can be found by:

$$f = \frac{D_p}{q}$$

For low angles of attack, the value of f can be considered a constant, so:

$$D_p = fq = \frac{f\sigma V^2}{295}$$

Parasite drag varies directly as the velocity squared:

$$\frac{D_{P_2}}{D_{P_1}} = \left(\frac{V_2}{V_1}\right)^2$$

(5.6)

Figure 5.13 shows a plot of parasite drag variation with velocity.

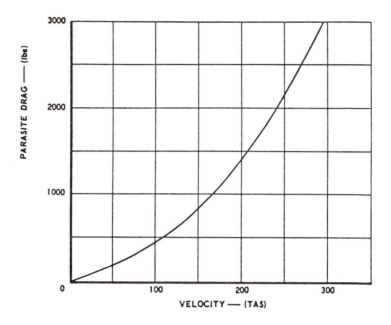

Figure 5.13   Parasite drag versus velocity.

TOTAL DRAG

Total drag, D_T, is simply the sum of the induced drag and the parasite drag:

$$D_T = D_i + D_p$$

As we mentioned earlier, the drag is simply the component of the aerodynamic force that is parallel to the free stream RW. It is made up of both the induced drag vector and the parasite drag vector, as shown in Figure 5.14.

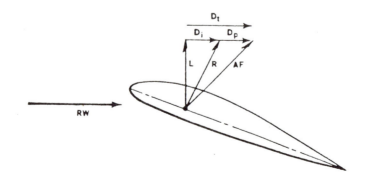

Figure 5.14   Drag vector diagram.

The total drag curve for an airplane with an induced drag curve as shown in Figure 5.11 and a parasite drag curve as shown in Figure 5.13 is shown in Figure 5.15.

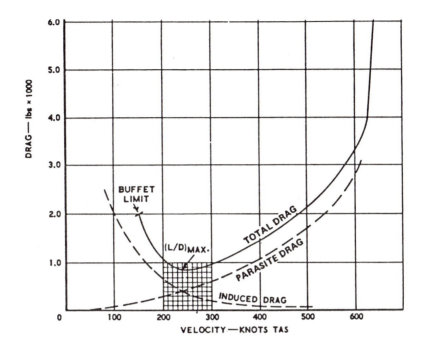

Figure 5.15   Total drag curve.

Note that the minimum drag or $(L/D)_{MAX}$ point occurs where the induced drag and parasite drag curves intersect.

Therefore, at this velocity, $D_i = D_p = \frac{1}{2}D_T$.

An equation for calculating total drag can be derived by combining equations 5.5 and 5.6.

Let condition 1 be at $(L/D)_{MAX}$ and condition 2 be any other point on the curve.

Remember that $D_{i_1} = D_{p_1} = \frac{1}{2}D_{MIN}$, as explained above.

$$D_T = \frac{1}{2}D_{MIN}\left\{\left(\frac{V_1}{V_2}\right)^2 + \left(\frac{V_2}{V_1}\right)^2\right\}$$

(5.7)

The limitations of this curve must be emphasized. At the low speed region, the high AOA required results in more "equivalent parasite area" than was calculated. The value of f is not constant. The value of parasite and total drag will be greater than the equation indicates. In the high speed region, faster airplanes may encounter compressibility effects and resultant "wave drag." This, too, has not been taken into account in equation 5.7.

The total drag curves, as described, are not accurate at an airspeed below about $1.3\,V_S$ or above about 0.7 Mach.

Drag

SYMBOLS AND UNITS

English Symbols

AR	Aspect ratio
b	Wing span (ft)
C	Chord (ft)
C_{AV}	Average chord
C_T	Tip chord
D_i	Induced drag (lb)
D_p	Parasite drag
D_T	Total drag
f	Equivalent parasite area (ft^2)
$(L/D)_{MAX}$	Maximum lift/drag ratio
MAC	Mean aerodynamic chord (ft)
2w	Downwash velocity (kts)

Greek Symbols

α_o	AOA of infinite wing
α_i	Induced AOA of finite wing
λ	(lambda) Taper ratio
Λ	(LAMBDA) Sweep angle (degrees)

EQUATIONS

5.1 $D = C_D qS$

5.2 $L/D = C_L/C_D$

5.3 $D_{MIN} = \dfrac{W}{(L/D)_{MAX}}$

5.4 $\alpha = \alpha_o + \alpha_i$

5.5 $\dfrac{D_{i_2}}{D_{i_1}} = \left(\dfrac{V_1}{V_2}\right)^2$

5.6 $\dfrac{D_{p_2}}{D_{p_1}} = \left(\dfrac{V_2}{V_1}\right)^2$

5.7 $D_T = \dfrac{1}{2}D_{MIN}\left\{\left(\dfrac{V_1}{V_2}\right)^2 + \left(\dfrac{V_2}{V_1}\right)^2\right\}$

76

PROBLEMS

1. The coefficient of drag can be defined as:

    a. the ratio of lift forces to drag forces.
    b. a measure of the efficiency of the airplane.
    c. the ratio of drag forces to lift forces.
    d. the ratio of drag pressure to dynamic pressure.

2. Laminar flow airfoils have less drag than conventional airfoils:

    a. in the high C_L region.
    b. in the high AOA region.
    c. in the high speed region.
    d. in the landing phase of flight.

3. Laminar flow airfoils have less drag than conventional airfoils because:

    a. the adverse pressure gradient starts farther back on the airfoil.
    b. the airfoil is thinner.
    c. more of the airflow is laminar.
    d. both a. and c. above.

4. Lift/Drag ratio is:

    a. a measure of the aircraft's efficiency.
    b. a maximum when the drag is a minimum.
    c. numerically equal to the glide ratio.
    d. all of the above.

5. An airplane flying at $C_{L(MAX)}$ will have:

    a. more parasite drag than induced drag.
    b. more induced drag than parasite drag.
    c. equal amounts of parasite and induced drag.
    d. not enough information given to determine.

6. Induced drag is:

    a. more important to low aspect ratio airplanes than to high aspect ratio airplanes.
    b. reduced when the airplane enters "ground effect".
    c. reduced if the airplane has winglets.
    d. all of the above.

Drag

7. Induced drag results from the lift vector being tilted to the rear. This is caused by:

    a. the tip vortices cause downwash behind the wing.
    b. the relative wind behind the wing is pushed downwards.
    c. the local relative wind at the AC is depressed.
    d. all of the above.

8. An airplane with a heavy load _____ when lightly loaded.

    a. can glide farther than
    b. can not glide as far as
    c. can glide the same distance as

9. If an airplane has a symmetrical wing which has an angle of incidence of 0° during takeoff, all the drag is:

    a. profile drag.
    b. parasite drag.
    c. wave drag.
    d. induced drag.

10. A low tailed airplane with static ports beneath the wing leaves "ground effect" after takeoff. It will experience:

    a. increased drag, nose up pitch, lowered IAS.
    b. increased drag, nose up pitch, higher IAS.
    c. decreased drag, nose down pitch, lowered IAS.
    d. decreased drag, nose up pitch, higher IAS.

11. A 10,000 lb aircraft has the C_L, C_D, and L/D curves shown in Figure 5.3. If $(L/D)_{MAX}$ occurs at 5° AOA and its value is 9.8, find the minimum drag on the aircraft.

12. If the above aircraft has a wing area S = 200 ft^2 and the value of C_L = 0.5 at $(L/D)_{MAX}$, find the sea level airspeed where D_{MIN} occurs.

13. For the above aircraft find D_i and D_p at $(L/D)_{MAX}$.

14. For the above aircraft complete the following table and plot D_i, D_p, and D_T on a sheet of graph paper.

V_2	$(V_2/V_1)^2$	$D_p = \frac{1}{2}D_{MIN}(V_2/V_1)^2$	$(V_1/V_2)^2$	$D_i = \frac{1}{2}D_{MIN}(V_1/V_2)^2$	$D_T = D_p + D_i$
125					
150					
172					
200					
300					
400					

CHAPTER SIX

JET AIRCRAFT BASIC PERFORMANCE

The performance capabilities of an aircraft depend on the relationship of the forces acting on it. The principal forces are lift, weight, thrust, and drag. If these forces are in equilibrium, as shown in Figure 6.1, the aircraft will maintain steady velocity, constant altitude flight. If any of the forces acting on the aircraft change, the performance of the aircraft will also change. To better understand the relationship between these forces and performance, we will study aircraft performance curves in this chapter. Some of the items of performance that can be predicted from these curves are:

1. Maximum level flight velocity.

2. Maximum climb angle.

3. Velocity for maximum climb angle.

4. Maximum rate of climb.

5. Velocity for maximum rate of climb.

6. Velocity for maximum endurance.

7. Velocity for maximum range.

This chapter shows the construction and use of the performance curves for THRUST PRODUCING AIRCRAFT. By thrust producing aircraft we are referring to turbojet and turbofan aircraft. Propeller and rotor driven aircraft are called POWER PRODUCING AIRCRAFT and will be discussed in Chapter Eight.

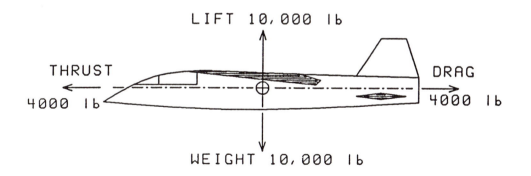

Figure 6.1 Aircraft in flight in equilibrium.

THRUST-PRODUCING AIRCRAFT

Turbojet, fanjet, ramjet, and rocket-driven aircraft produce thrust directly from their engines and are called thrust producers. Fuel consumption is roughly proportional to the thrust output of these aircraft and, because range and endurance performance are functions of fuel consumption, the thrust required to fly is of prime interest.

THRUST-REQUIRED CURVE

Each pound of drag requires a pound of thrust to offset it. In Chapter Five we discussed how the drag curve for an airplane is drawn. We can now call this same curve a thrust-required curve. The thrust-required, T_R, curves in this chapter are for the T-38A supersonic turbojet trainer. Because this aircraft encounters a large amount of "wave" drag at high speeds, a sharp increase in thrust required will be noted above 600 knots.

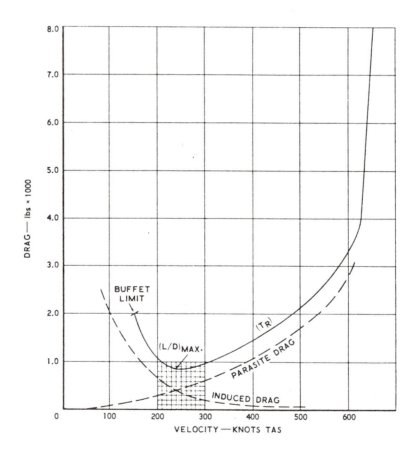

Figure 6.2   T-38 thrust required (drag) curve.

Figure 6.2 shows the drag or thrust required curve for a 10,000lb. T-38 in the clean configuration, at sea level on a standard day. This aircraft has a drag of 2000 lb. at the buffet limit (stall speed) of about 150 knots. At about 485 knots the drag is also 2000 lb. Minimum drag (or minimum T_R) occurs at $(L/D)_{MAX}$, which is at 240 knots for this weight aircraft, and has a value of 830 lb.

PRINCIPLES OF PROPULSION

All aircraft powerplants have certain common principles. These are based on Newton's second and third laws. You will recall that the second law can be written as $F = ma$. Simply stated, an unbalanced force, F, acting on a mass, m, will accelerate, a, the mass in the direction of the force.

The force is provided by the expansion of the burning gases in the engine.

The mass is the mass of the air passing through the jet engine or through the propeller/rotor system.

The acceleration is the change in the velocity of the intake air per unit of time.

Newton's third law states that for every action force, there is an equal and opposite reaction force. It is this reaction force that provides the *thrust available*, (T_A), to propel the aircraft.

Figure 6.3 shows a schematic of the process of producing thrust in a jet-type engine.

Figure 6.3   Engine thrust schematic.

$$T_A = Q(V_2 - V_1) \tag{6.1}$$

where

T_A = thrust available (lb).

Q = mass airflow = ρAV (slugs/sec; equation 2.6)

V_1 = inlet (flight) velocity (fps)

V_2 = exit velocity (fps)

Equation 6.1 means that the thrust output of an engine can be increased by either (1) increasing the mass airflow or (2) increasing the exit velocity/flight velocity, (V_2/V_1) ratio.

The first alternative requires that the cross sectional area of the engine be increased to allow more air to be processed.

The second alternative requires increasing the relative velocity, $V_2 - V_1$. This imparts higher kinetic energy to the gases. The kinetic energy is wasted in the exhaust airstream, resulting in decreased efficiency of the engine.

Propulsive efficiency, η_p, can be expressed by:

$$\eta_p = \frac{2V_1}{V_2 + V_1}$$

(6.2)

As V_2 increases, propulsive efficiency goes down. Early turbojet engines were used on jet fighters. To keep the drag of the engines within reason, the engines could not have a large frontal area. This limited the amount of air that could be processed, and high exhaust velocities were required to obtain the needed thrust. As we have seen, this meant poor efficiency for these engines. As larger aircraft were developed, it was possible to use engines with larger frontal area. The fanjet engines were developed, and the required thrust was produced by increasing Q, rather than by increasing V_2, thus improving the efficiency of the engines.

THRUST-AVAILABLE TURBOJET AIRCRAFT

Turbojet aircraft engines are rated in terms of static thrust. The aircraft is restrained from moving, and the thrust is measured and converted to standard sea level conditions.

It can be seen from equation 6.1,

$$T_A = Q(V_2 - V_1) = \rho AV_1(V_2 - V_1),$$

that as the aircraft gains velocity, (V_1), the mass airflow increases, but the acceleration through the engine decreases (V_2 is nearly constant). The thrust available is nearly constant with airspeed and is considered to be a constant in this discussion.

Thrust varies, however, if the RPM of the engine is changed. The change in thrust is not linear with RPM. For example a reduction of RPM of one percent from 100 percent may result in a loss of five percent thrust. Modern jet engines are equipped with *engine pressure ratio*, EPR, gages. These give a more accurate indication of the thrust being developed.

Thrust is reduced with an increase in altitude. Equation 6.1 indicates that the thrust decreases as the density of the air decreases. There is another factor, temperature, that is also involved, but does not appear in the equation. Lower temperatures at altitude improve efficiency, so the loss in thrust is not as great as the decrease in density.

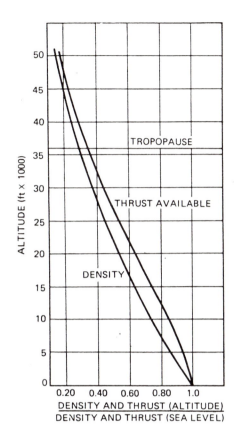

Figure 6.4   Thrust variation with altitude.

SPECIFIC FUEL CONSUMPTION

Specific fuel consumption, c_T, is the fuel flow per pound of thrust developed by a jet engine.

$$c_T = \frac{\text{fuel flow (lb./hr.)}}{\text{thrust (lb.)}}$$

Specific fuel consumption is a measure of the efficiency of an engine. Low values of c_T are desirable.

Turbojet engines are designed to operate at high RPM and will produce low specific fuel consumption values in this region. Figure 6.5 shows the variation of c_T with RPM.

The beneficial effect of lower temperature at altitude also results in decreased specific fuel consumption, as shown in Figure 6.6. Above the tropopause the temperature is constant, so no further reduction in c_T will occur. In fact, the compressor will not operate efficiently because of the low air density, and an increase in c_T will result.

83

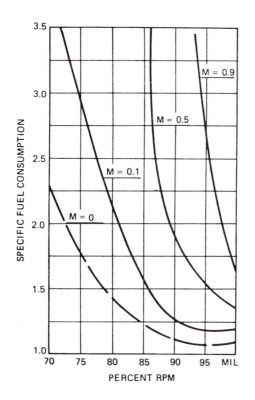

Figure 6.5   Specific fuel consumption vs. RPM.

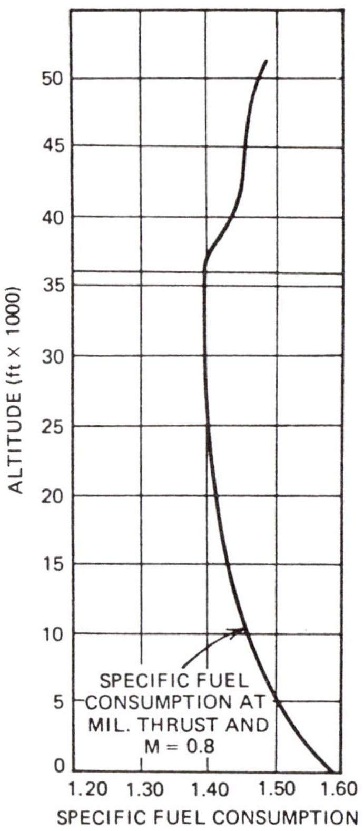

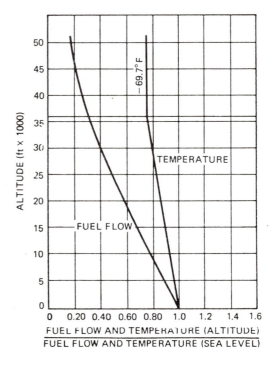

Figure 6.6   Specific fuel consumption vs. altitude.

FUEL FLOW

The total fuel flow, FF, equals the thrust of the engine multiplied by the specific fuel consumption:

$$FF = T(c_T)$$

Thrust varies with altitude as shown in Figure 6.4 and c_T varies with altitude as shown in Figure 6.6.

The fuel flow at altitude, as a ratio of that at sea level, at a fixed RPM (100 percent) and airspeed (MACH = 0.8) is shown in Figure 6.7.

Figure 6.7   Fuel flow vs. Altitude.

THRUST AVAILABLE-THRUST REQUIRED CURVES

Performance depends on the relationship of the thrust-available and thrust- required curves. If the thrust available is equal to the thrust required, the aircraft can fly straight and level, but cannot climb or accelerate (without losing altitude). An excess of T_A over T_R is required for these maneuvers.

The thrust-available and thrust-required curves are plotted for the T-38 at 10,000 lb. gross weight, clean configuration, sea level, in Figure 6.8.

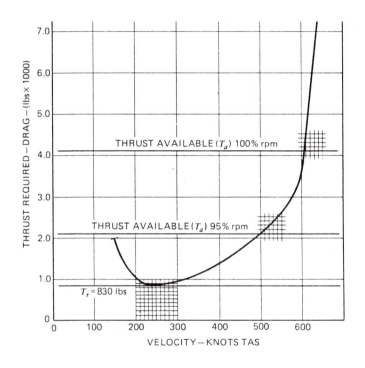

Figure 6.8   T-38 thrust available versus thrust required.

The thrust available is plotted for two different conditions (100 and 95 percent RPM). This reduction in T_A with RPM was discussed previously. These curves show that a 5 percent reduction in RPM results in a reduction in T_A from 4100 to 2100 lb.

This aircraft could not maintain altitude at any speed if the thrust is reduced below 830 lb.

Next we will show how the curves are used to predict the performance items that we listed at the first of this chapter. As the curves that we have shown so far are for one weight, clean configuration, and at sea level, our calculations are for these conditions only.

In the next chapter we will show how weight, configuration, and altitude change the curves.

ITEMS OF AIRCRAFT PERFORMANCE

Straight and Level Flight

When an aircraft is in steady (unaccelerated) level flight, it must be in equilibrium. Lift must be exactly equal to the weight of the aircraft, and the thrust on the aircraft must be exactly equal to the drag. This second condition exists when the thrust available is equal to the thrust required.

This happens when the T_A and T_R curves intersect. There are two possibilities for such an intersection. One is at the high speed region of flight; the other is at the low speed area. The problems in the low speed area are discussed in later chapters. The conditions for flight at maximum level velocity, V_{MAX}, are discussed here.

For non-afterburner flight the thrust available for the T-38 is 4100 lb. From Figure 6.8 it can be seen that the high speed intersection point of T_A and T_R is at 605 knots true airspeed. With lesser thrust settings the V_{MAX} will be less. For 95 percent RPM the value of V_{MAX} is 500 knots TAS.

Climb Performance

Every flight requires that the pilot climb the aircraft to some altitude. Mission requirements dictate the climb schedule. For instance, an intercept mission would require that altitude be gained as rapidly as possible, while a mission of long range would require that much distance be covered in the climb.

There are two basic types of climb.

One type involves a delayed climb, in which the pilot builds up airspeed (kinetic energy) before starting the climb. He then "zooms" the aircraft to altitude. This converts the kinetic energy to potential energy (height). This maneuver is used mostly by fighter aircraft pilots in intercept tactics, setting climb to altitude records, and other maneuvers.

The other type of climb is the steady velocity climb. It is used much more often and will be discussed here. This type of climb is an equilibrium condition, with all the forces along the flight path being balanced. This is shown in Figure 6.9.

The forces acting along the flight path are thrust, T, and drag, D. Lift acts 90 degrees to the flight path and the weight, W, acts toward the center of the earth. The climb angle is designated by the Greek letter gamma, γ.

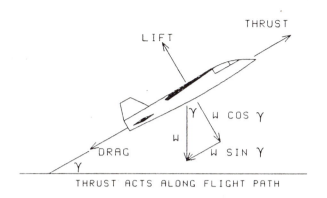

Figure 6.9   Forces acting on a climbing aircraft.

The weight vector is replaced by two other vectors. One is perpendicular to the flight path (W cos γ) and the other is parallel to the flight path (W sin γ). It should be noted that W sin γ actually acts at the aircraft's CG, not as shown on the figure. For steady velocity the forces along the flight path must be balanced. Thus an equation can be written as:

$$T - D - W \sin \gamma = 0 \tag{6.3}$$

Angle Of Climb

Equation 6.3 can be rewritten as:

$$\sin \gamma = \frac{T - D}{W} = \frac{T_A - T_R}{W} \tag{6.3a}$$

As an angle increases in size, the sine of the angle increases in value. Therefore the maximum climb angle will occur at $(T_A - T_R)_{MAX}$, where maximum excess thrust exists. For a turbojet, with the thrust available assumed to be constant with velocity, this will be where T_R (drag) is a minimum. This, of course, is at $(L/D)_{MAX}$. This is shown in Figure 6.10.

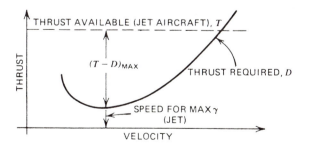

Figure 6.10   Velocity for maximum climb angle.

When an aircraft is in steady flight, horizontally, the lift is equal to the weight of the aircraft. Figure 6.9, however, shows that the lift is less than the weight if the aircraft is in a climb. This is possible because some of the thrust supports some of the weight of the aircraft. The steeper the climb is, the less the lift supports the weight. The extreme case occurs when the aircraft climbs straight up ($\gamma = 90°$). Then the lift would be zero, and the thrust would support the entire weight of the aircraft and also overcome the drag. Obviously a thrust/weight ratio of greater than 1 is required for this maneuver.

Obstacle clearance is affected by headwind and tailwind as shown in Figure 6.11.

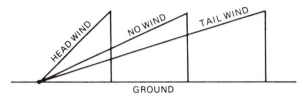

Figure 6.11   Wind effect on climb angle to the ground.

Once the airplane is airborne, the groundspeed is reduced by a headwind and increased by a tailwind. With reduced groundspeed, the time to reach the obstacle is increased, and more altitude is gained.

One other factor in obstacle clearance is the time required to reach the $(L/D)_{MAX}$ velocity. After the jet takes off, it must accelerate to this velocity and, in doing so, precious ground distance is being used up. Even though the jet is finally climbing at its highest climb angle, it may not clear the obstacle. The aircraft may clear the obstacle if the climb is started sooner, at a lower speed and at a lesser climb angle. This is shown in Figure 6.12.

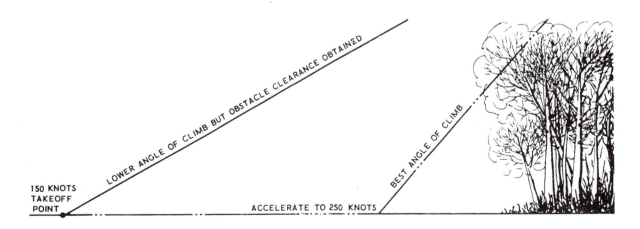

Figure 6.12   Obstacle clearance for jet takeoff.

Rate Of Climb

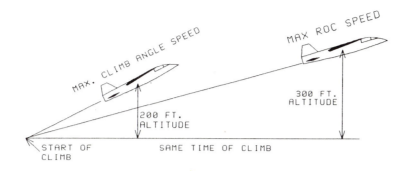

Figure 6.13   Climb angle and rate of climb.

The aircraft on the left is flying at maximum climb angle. The airspeed of this aircraft is relatively slow, compared to the aircraft on the right. The aircraft on the right is flying at a smaller climb angle, but the much higher airspeed more than compensates for the lower angle, and it is at a higher altitude. The aircraft on the right has a higher rate of climb, RC.

Figure 6.14 is a vector diagram relating rate of climb to airspeed and climb angle.

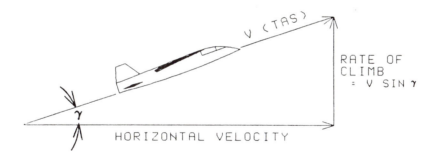

Figure 6.14   Rate of climb velocity vector.

We can solve the right triangle in Figure 6.14 for the vertical component of the flight velocity. The rate of climb is then:

$$RC = V_K \sin \gamma \quad \text{(knots)}$$

Substituting the value of sin γ from equation 6.3a

$$RC = (V_K)\left(\frac{T_A - T_R}{W}\right) \quad \text{(knots)}$$

$$RC = 101.3 \, (V_K)\left(\frac{T_A - T_R}{W}\right) \quad \text{(fpm)} \tag{6.4}$$

The velocity for maximum rate of climb cannot be determined by a simple examination of the T_A-T_R curves as was done to find the velocity for maximum angle of climb. Equation 6.4 shows that RC depends upon velocity **and** excess thrust, both of which are variables. However, by selecting various airspeeds and finding the corresponding values of T_A and T_R from Figure 6.8, then solving equation 6.4, a plot of RC versus velocity can be made similar to Figure 6.15 and the velocity for maximum RC can be found.

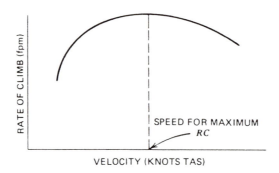

Figure 6.15   Velocity for maximum rate of climb.

ENDURANCE

The *endurance* of an aircraft is the amount of time that it can remain airborne. Maximum endurance is independent of distance covered. It is useful when holding due to weather, air traffic etc. Endurance is inversely proportional to fuel flow. The lower the fuel flow, the greater the endurance.

Figure 6.16 shows the thrust-required curve for the T-38 with the fuel flow also plotted on the vertical scale. Note that the minimum fuel flow occurs at the minimum thrust required. This, again, is at $(L/D)_{MAX}$.

Maximum endurance is another of the items of aircraft performance for a jet aircraft that occurs at $(L/D)_{MAX}$.

For jet aircraft, we have seen that the following items of performance occur at $(L/D)_{MAX}$:

1. Minimum drag.

2. Maximum engine-out glide range.

3. Maximum climb angle.

4. Maximum endurance.

It should be obvious that it is important to know the airspeed or AOA for $(L/D)_{MAX}$. Airspeed will vary with aircraft weight; AOA will not. Indicated airspeed for $(L/D)_{MAX}$ will not vary with altitude.

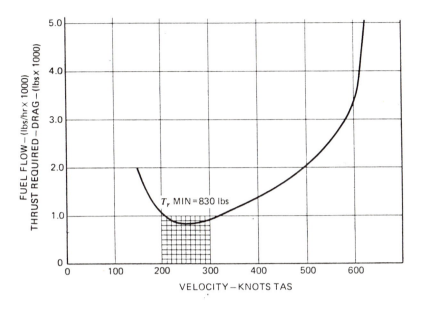

Figure 6.16   Finding maximum endurance velocity.

Specific Range

To get maximum range, an aircraft must fly the maximum distance with the fuel available, thus the specific range, SR, must be a maximum. Specific range can be defined by the following relationship:

$$\text{Specific range} = \frac{\text{nmi}}{\text{lb. fuel}}$$

$$= \frac{\text{nmi/hr.}}{\text{lb. fuel/hr.}}$$

$$= \frac{\text{speed (knots)}}{\text{fuel flow (lb./hr.)}}$$

$$(SR)_{MAX} = \left(\frac{V}{FF}\right)_{MAX}$$

Which means

$$(SR)_{MAX} \text{ occurs at } \left(\frac{FF}{V}\right)_{MIN}$$

91

Graphically, the maximum specific range point can be shown on a fuel flow versus velocity curve, such as is shown in Figure 6.17.

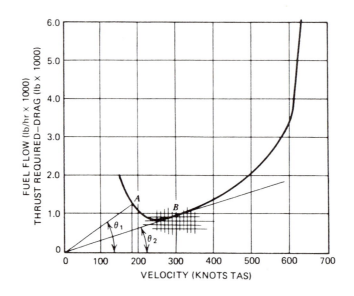

Figure 6.17   Finding the maximum specific range velocity.

Select any point on the T_R curve, such as A. Draw a straight line from the point to the origin and also drop a vertical line from the point to the velocity scale. Label the angle at the origin of the resulting right triangle, angle θ_1. The tangent of this angle equals the opposite side of the triangle divided by the adjacent side. The opposite side is a measure of the fuel flow, and the adjacent side of the triangle is the aircraft's velocity.

$$\tan \theta_1 = \frac{FF}{V}$$

Hence, if $\dfrac{FF}{V}$ is a minimum, $\tan \theta$ must be a minimum, which means that θ must be a minimum. Only one point on the curve meets this requirement. That is the point where the line from the origin is tangent to the T_R curve: at point B.

θ_2 is a minimum at B, and maximum specific range will occur if the aircraft is flown at this speed.

Maximum specific range occurs at a speed corresponding to the tangent point of a straight line drawn from the origin to the curve. This is not $(L/D)_{MAX}$, but is the point where $\left(\dfrac{\sqrt{C_L}}{C_D}\right)$ is a maximum.

92

Wind Effect on Specific Range

An aircraft flying into a headwind is in an unfavorable environment and thus should fly at a higher true airspeed to reduce the effect of the headwind. Likewise, an aircraft flying with a tailwind is being helped on its way and thus should slow down to take advantage of the wind. How much the aircraft's airspeed should be altered is shown in Figure 6.18.

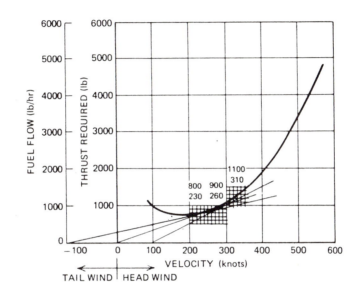

Figure 6.18   Wind effect on specific range.

This illustration shows how a 100 knot wind will affect the airspeed for maximum range. To find the corrected airspeed for a 100 knot headwind, plot the headwind on the velocity scale to the right of the origin. Use this point as a new origin and draw a tangent line to the curve. This will give the new airspeed for best range. In the example the no wind airspeed for maximum range was 260 knots and the corrected airspeed is 310 knots.

Correcting for a tailwind is done in a similar manner. The tailwind is plotted on the velocity scale, but to the left of the origin. A new tangent line will show a corrected airspeed of 230 knots.

Total Range

The total range of an aircraft depends on the fuel available and the specific range. Specific range is a variable. As the weight of an aircraft decreases with fuel consumption, the specific range improves. This variation is evident from Figure 6.19.

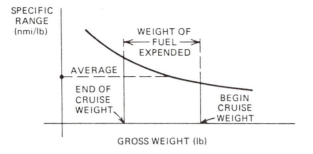

Figure 6.19 Total range calculation

The area under the curve, between the beginning and ending cruise weights, is a measure of the total range. An easy way of approximating the total range is to find an average specific range and multiply this by the total fuel consumed.

SYMBOLS AND UNITS

English Symbols

c_T	Specific fuel consumption (lb fuel/hr/lb thrust)
FF	Fuel Flow (lb/hr)
Q	Mass airflow (slugs/sec)
RC	Rate of climb (fpm)
SR	Specific Range (nmi/lb fuel)
T_A	Thrust available (lb)
T_R	Thrust required
V_1	Intake (flight) velocity (fps)
V_2	Exit velocity
V_K	Aircraft velocity (knots)

Greek Symbols

γ (gamma) Climb angle (degrees)
η_p (eta)   Propulsion efficiency (%/100)

EQUATIONS

6.1 $T = Q(V_2 - V_1)$

6.2 $\eta_p = \dfrac{2V_1}{V_2 + V_1}$

6.3 $\sin \gamma = \dfrac{T - D}{W} = \dfrac{T_A - T_R}{W}$

6.4 $RC = 101.3(V_K)\left(\dfrac{T - D}{W}\right)$ (fpm)

PROBLEMS

1. The equation $T_A = Q(V_2-V_1)$ shows:

    a. you can get more thrust from a jet engine by using water injection.
    b. you can get more thrust from a jet engine by decreasing the exhaust pipe area.
    c. neither of the above.
    d. both a. and b. above.

2. Q in $T_A = Q(V_2-V_1)$ is the mass flow (slugs per sec.).

  The mass flow depends on:

    a. the cross sectional area of the turbojet engine inlet.
    b. the engine inlet velocity.
    c. the density of the air at the same point.
    d. all of the above.

3. The equation $\eta_P = \dfrac{2V_1}{V_2 + V_1}$ shows that decreasing the exhaust pipe area will:

    a. decrease the propulsive efficiency of a jet engine.
    b. increase the propulsive efficiency.
    c. increase the thrust of the engine.
    d. decrease the thrust of it.

4. Specific fuel consumption of a turbine engine at 35,000 feet altitude compared to that at S.L. is:

    a. less
    b. more
    c. the same

5. The thrust available from a jet engine at 35,000 feet altitude compared to that at S.L. is:

    a. less
    b. more
    c. the same

6. Fuel flow for a jet at 100 percent RPM at altitude compared to 100 percent RPM at sea level is:

    a. less
    b. more
    c. the same

7. A pilot is flying his airplane at the speed for best range under no wind conditions. A tailwind is encountered. To get best range now he must:

    a. speed up by an amount less than the wind speed.
    b. slow down by an amount equal to the wind speed.
    c. slow down by an amount more than the wind speed.
    d. slow down by an amount less than the wind speed.

8. To obtain maximum range a jet airplane must be flown at:

    a. a speed less than that for $(L/D)_{MAX}$.
    b. a speed equal to that for $(L/D)_{MAX}$.
    c. a speed greater than that for $(L/D)_{MAX}$.

9. Maximum rate of climb for a jet airplane occurs at:

    a. a speed less than that for $(L/D)_{MAX}$.
    b. a speed equal to that for $(L/D)_{MAX}$.
    c. a speed greater than that for $(L/D)_{MAX}$.

10. Maximum climb angle for a jet aircraft occurs at:

    a. a speed less than that for $(L/D)_{MAX}$.
    b. a speed equal to that for $(L/D)_{MAX}$.
    c. a speed greater than that for $(L/D)_{MAX}$.

11. A single engine turbojet aircraft is flying at 296 knots TAS at sea level. The mass flow rate through the engine is 10 slugs/sec. The exit velocity from the engine is 800 fps. Find: (a) the thrust of the engine and (b) the propulsive efficiency.

12. A jet airplane has thrust available as shown in Figure 6.8 and specific fuel consumption as shown in Figure 6.5. Find the fuel flow at sea level (military RPM and M = 0.7).

13. Using Figure 6.7 find the fuel flow for the airplane in problem 12 at the tropopause.

14. A jet airplane has the T_A-T_R curves shown in Figure 6.8. The airplane data: weight = 10,000 lb.; standard sea level day; clean configuration. Find:

    a. V_{MAX} at 95% RPM;
    b. V for maximum climb angle (100% RPM);
    c. sine of maximum climb angle (100% RPM);
    d. rate of climb at 400 knots (100% RPM);
    e. V for best endurance;
    f. V for maximum range;
    g. V for maximum range with 100 knots of headwind.

CHAPTER SEVEN

JET AIRCRAFT APPLIED PERFORMANCE

The aircraft performance items that we discussed in Chapter Six were for one weight, clean configuration, and sea level standard day conditions. In this chapter we will alter these conditions and examine the resulting performance. To simplify the explanation, only one condition will be altered at a time.

VARIATIONS IN THE THRUST-REQUIRED CURVE

The basic thrust required curve for the T-38 (Figure 6.2) was drawn for a 10,000 lb airplane in the clean configuration, at sea level on a standard day. We will now show how the curve changes for variations in weight, configuration, and altitude.

Weight Changes

Changing the weight of an aircraft affects the induced drag much more than the parasite drag. The only change in the parasite drag is caused by small changes in equivalent parasite area, f, at varying angles of attack. Only changes in induced drag are considered here. The change in the induced drag of the T-38 at 8000 and at 12,000 lb. gross weight is shown in Figure 7.1.

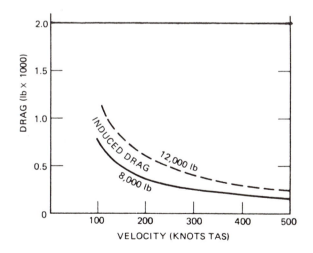

Figure 7.1 Effect of weight change on induced drag.

The total drag is also increased by the addition of weight. The intersection of the induced drag and the parasite drag curves still marks the point of $(L/D)_{MAX}$, but it is moved up and to the right. In fact, all points on the thrust-required curve are moved up and to the right. Because induced drag is greater than parasite drag in the low speed region of flight, the curve is moved by a greater amount in the low speed region than in the high speed region. The total drag curves are shown as the thrust required curves in Figure 7.2.

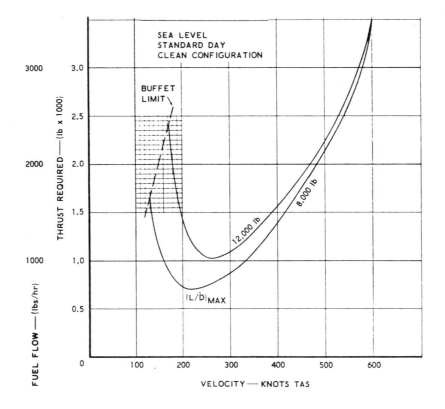

Figure 7.2   Effect of weight change on T_R curve.

Both of the curves shown above were obtained by altering the basic T_R curve for the 10,000 lb. T-38 (Figure 6.2). First the airspeed for each point on the curve was corrected by using equation 4.2 and, second the drag was corrected by using the principle that (L/D) is not affected by weight.

This was discussed on page 63 where it was stated, "Note that the value of $(L/D)_{MAX}$ and the angle of attack at which it occurs do not vary with aircraft weight or altitude." Not only does $(L/D)_{MAX}$ not vary but other (L/D) values do not vary either. They are functions of AOA only. So, if (L/D) does not vary with weight, then if weight is changed, drag is changed proportionately and:

$$\frac{L_2}{D_2} = \frac{L_1}{D_1}$$

98

Under 1 G flight L = W, so

$$\frac{D_2}{D_1} = \frac{W_2}{W_1} \tag{7.1}$$

Applying equations 4.2 and 7.1 to selected airspeed and corresponding drag, (T_R), values of Figure 6.2 and plotting produces Figure 7.2.

Later in this chapter we will discuss how the pilot must use this curve to obtain maximum performance from his aircraft as the weight changes.

Configuration Changes

When a pilot lowers his landing gear (and/or flaps), the equivalent parasite area is greatly increased. This increases the parasite drag of the aircraft, as shown in Figure 7.3, but it has little effect on the induced drag.

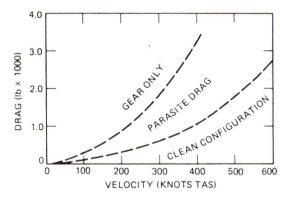

Figure 7.3   Effect of configuration on parasite drag

The effect on the thrust-required curve is to move it upward and to the left. The amount that the curve moves is not easily calculated. The equivalent parasite area must be known for the aircraft in the dirty configuration, and the new parasite drag values must be calculated. The T_R curve can then be found by adding the new D_p curve values to the old D_i curve values and plotting.

The thrust-required curves for several configurations of the T-38 is shown in Figure 7.4. In discussing the effect of weight change we saw that the values of (L/D) remained constant at the same AOA. The value of $(L/D)_{MAX}$ does not change as weight varies. However, this certainly is not true for a configuration change.

For the clean configuration, the value of $(L/D)_{MAX}$ in Figure 7.4 is 12.05. In the full flaps and gear configuration it is only 4.26.

99

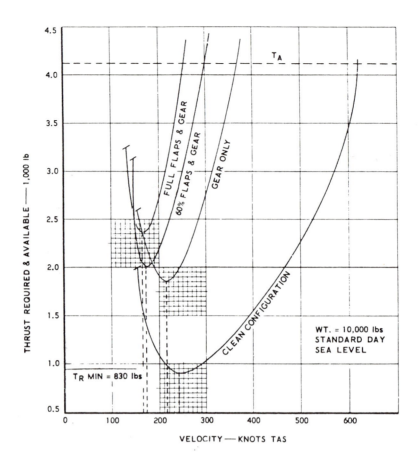

Figure 7.4 *Effect of configuration on T_R curve.*

Altitude Changes

Drag on an airplane depends on the dynamic pressure:

$$D = C_D q S$$

At any given AOA, the value of the drag coefficient, C_D, is a constant, and the wing area, S, is also a constant. Therefore the drag will also be a constant if the dynamic pressure is a constant. If we plotted drag versus equivalent airspeed, there would be only one drag curve for all altitudes. We could say that "drag at altitude equals drag at sea level," but we would have to qualify that statement by adding "at the same equivalent airspeed." Because the curves are drawn for true airspeed, different curves must be drawn for each altitude.

If the drag does not change, but only the true airspeed at which it occurs changes, the only correction we must make is one for airspeed:

$$\frac{V_2}{V_1} = \sqrt{\frac{\sigma_1}{\sigma_2}}$$

$$(7.2)$$

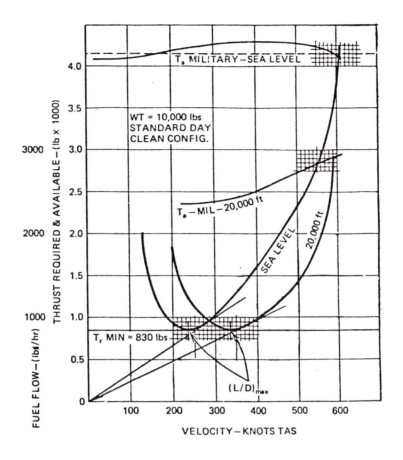

Figure 7.5   Effect of altitude on T_R and T_A curves.

The above curves show that the minimum thrust required is 830 lb. for both altitudes. Of course, this means that the value of $(L/D)_{MAX}$ is not affected by altitude. The True Air Speed for minimum drag does change as shown in equation 7.2.

If V_1 is at sea level, $\sigma_1 = 1.0$. and equation 7.2 is:

$$V_2 = \frac{V_1}{\sqrt{\sigma_2}}$$

In addition to the thrust-required curves, Figure 7.5 also shows thrust- available at sea level and 20,000 ft. The thrust reduction is in accordance with Figure 6.4.

VARIATIONS OF AIRCRAFT PERFORMANCE

Straight and Level Flight

Weight Change. Increase in weight means an increase in drag, but T_A is not affected, so V_{MAX} is decreased. A weight decrease will reduce the drag and the T_A-T_R intersection V_{MAX} will be at a higher airspeed.

Configuration Change. V_{MAX} is usually limited by structural strength of landing gear and flaps, therefore V_{MAX} is reduced by a dirty configuration.

Altitude Change. V_{MAX} true airspeed is only slightly affected by altitude (see Figure 7.5). The T_R curve moves to the right with an increase in altitude and the T_A decreases. V_{MAX} EAS will be less at altitude as T_A decreases.

Climb Performance

Angle of Climb

Weight Change. Increased weight means increased drag, while T_A is not changed. Result is that the (T – D) term in equation 6.3 is reduced and a lower angle of climb results. When weight is reduced the opposite occurs and climb angle is increased.

Configuration Change. Drag increases when gear and flaps lowered, so a smaller angle of climb results. Clean up the aircraft as soon as practical.

Altitude Increase. Thrust available decreases with increase in altitude, drag remains the same, so (T – D) is smaller and angle of climb is smaller at altitude.

Rate of Climb

Weight Change. Rate of climb is the velocity times the sine of the climb angle. As we saw above the (T – D) term reduces with a weight increase and the extra drag also slows the aircraft, so RC is reduced with increased weight and increased with decreased weight.

Configuration Change. A dirty aircraft reduces all climb performance, so clean up the aircraft as soon as practical.

Altitude Increase. T_R (drag) does not change with an increase in altitude but T_A is decreased, thus (T – D) is reduced. The velocity increases with altitude. The product of the two terms, however, decreases. When the T_A reduces to the point where it is tangent to the T_R curve, the absolute ceiling is reached and the RC is zero.

Endurance

Weight Change. More drag is developed with increased weight. Fuel flow is proportional to T_R (drag), so endurance is reduced, and vice versa.

Configuration Change. Pull up your gear and flaps to improve endurance. If gear or flaps cannot be retracted, slow down where parasite drag is less.

Altitude Increase. $T_{R(MIN)}$ is the same, but fuel flow is less with altitude so endurance is improved. A word of caution, you may burn more fuel climbing to a higher altitude than if you had remained at the lower altitude.

Specific Range

Weight Change. As fuel is burned, weight is reduced, and the thrust required curve moves downward and to the left. This is shown in Figure 7.6.

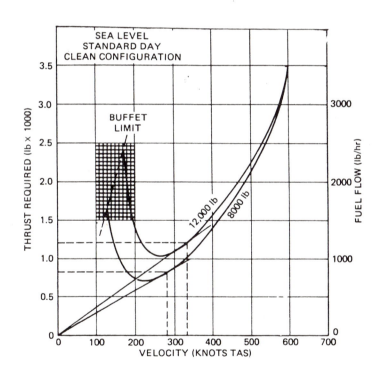

Figure 7.6   Effect of weight change on specific range.

The tangent lines, indicating velocity for $SR_{(MAX)}$, are shown in the figure. Two important changes in the flight schedule must be made by the pilot as weight decreases, if maximum range is to be obtained. First, the thrust of the engine must be reduced; and second, the speed of the aircraft must be reduced. However, a reduction in throttle will result in a reduction of thrust available, and this will automatically result in a lower airspeed. The natural tendency of the pilot to keep the throttle setting the same, as weight is reduced, and to allow the airspeed to increase must be suppressed. For maximum range, as fuel is burned, ***reduce airspeed.***

103

Configuration Change. Should it be necessary to fly in the dirty configuration, slow the airplane to the new tangent point to reduce parasite drag.

Altitude Increase. There are several advantages in flying jet aircraft at altitude. Lower temperatures increase engine efficiency. Reduced specific fuel consumption results from operating at high RPM. A third advantage can be seen from Figure 7.5. Substantially higher airspeeds at the tangent point to the T_R curve at altitude do not require higher fuel flow, thus the angle decreases indicating a better SR.

NOTE:

You should note that in our discussion of performance curves and items of performance for thrust-producing aircraft the word "power" was never used. Many pilots use the terms "power" and "thrust" interchangeably. This is a gross error.

Thrust-producing aircraft do not produce power. They do not use "power curves." Many times pilots will make statements such as "add power," "back side of the power curve," and others. All such statements are incorrect and should be avoided.

EQUATIONS

7.1 $\quad \dfrac{D_2}{D_1} = \dfrac{W_2}{W_1}$

7.2 $\quad V_2 = \dfrac{V_1}{\sqrt{\sigma_2}}$ (V_1 is at S.L. on Std. Day.)

PROBLEMS

1. If the weight of a jet airplane is increased the:

    a. Parasite drag increases more than induced drag.
    b. induced drag decreases more than parasite drag.
    c. both parasite and induced drag increase by the same amount.
    d. induced drag increases more than parasite drag.

2. If the weight of a jet airplane is reduced as fuel is burned, the T_R curve:

    a. moves down and to the right.
    b. moves up and to the right.
    c. moves down and to the left.
    d. moves up and to the left.

3. If a jet airplane is in the gear down configuration, the increase in:

    a. parasite drag is more than that of the induced drag.
    b. induced drag is more than that of the parasite drag.
    c. both types of drag are the same.

4. If it is impossible to raise the landing gear of a jet airplane, to obtain best range, the airspeed must be _____ from that for the clean configuration.

    a. increased.
    b. decreased.
    c. not changed.

5. From Figure 7.4 find the glide ratio $(L/D)_{MAX}$ for the airplane in the 60% flaps and gear down configuration.

    a. 12.05
    b. 5.4
    c. 5.0
    d. 4.2

6. The minimum drag for a jet airplane does not vary with altitude.

    a. True.
    b. False.

7. Figure 7.5 shows an increase in specific range with altitude because:

    a. T_R decreases while fuel flow decreases.
    b. T_R remains the same while fuel flow decreases.
    c. T_R remains the same while T_A decreases.
    d. Fuel flow remains about the same while airspeed increases.

8. A jet airplane is flying to obtain maximum specific range. As fuel is burned the pilot must:

    a. reduce throttle but maintain the same airspeed.
    b. maintain throttle setting and let the plane accelerate.
    c. reduce throttle and airspeed.
    d. maintain throttle and reduce airspeed.

9. A lightly loaded airplane will be able to glide farther but at a lower airspeed than when it was heavily loaded.

    a. True.
    b. False.

10. Which of the following words should never be used in the discussion of jet aircraft?

    a. Power.
    b. Horsepower.
    c. Power curve.
    d. All of the above.

11. Using Figure 7.2, find the velocity for best range for the airplane at 12,000 and 8000 lb. G.W.

12. Using Figure 7.2, calculate the specific range for the airplane at both weights if it is flying at the best range airspeed.

13. Using Figure 7.2, calculate the specific range of the above airplane, if the throttle is not retarded from the 12,000 lb. best range position, and the airspeed is allowed to increase as fuel is burned and weight is reduced to 8000 lb.

Compare your answer to that of problem 12 and state your conclusions.

14. Using Figure 7.5, calculate the specific range for this airplane if it is flying at best range airspeed at sea level and at 20,000 ft. altitude.

CHAPTER EIGHT

PROP AIRCRAFT BASIC PERFORMANCE

All aircraft in flight must produce thrust to overcome the drag of the aircraft. In turbojets and in other thrust producing aircraft this thrust is produced directly from the engine. In aircraft that have propellers (or rotors), the engine does not produce thrust directly. These aircraft are called *power producers* because the engine produces power, which turns the propeller. The propeller then develops an aerodynamic force as it turns through the air; this force is thrust.

Fuel consumption of power-producing aircraft is roughly proportional to the power produced, rather than to the thrust produced. Range and endurance performance are functions of fuel consumption, and so the power required to fly the aircraft is of prime importance.

Reviewing the relationship between thrust and power that was discussed in Chapter One, we find:

$$\text{Thrust} = \text{force} = \text{a push or pull (lb)}$$

$$\text{Work} = \text{force} \times \text{distance (ft-lb)}$$

$$\text{Power} = \frac{\text{work}}{\text{time}} = \frac{\text{force} \times \text{distance}}{\text{time}} \text{ (ft-lb/sec)}$$

But:

$$\frac{\text{distance}}{\text{time}} = \text{velocity}$$

so:

$$\text{Power} = \text{force} \times \text{velocity} = FV = TV \text{ (ft-lb/sec)}$$

$$\text{Horsepower} = \frac{\text{power}}{550} = \frac{TV_{FPS}}{550}$$

If V is in knots then, as shown in equation 1.6:

$$HP = \frac{TV_K}{325}$$

POWER-REQUIRED CURVES

Equation 1.6 allows us to convert the drag (T_R) curve into the horsepower required, P_R, curve.

Before we convert the total drag curve into a total power-required curve, let us see how the power required to overcome the induced drag, P_{Ri}, and the power required to overcome the parasite drag, P_{Rp}, are affected.

Equation 1.6 can be rewritten as:

$$P_R = \frac{DV_K}{325}$$

Selecting values of drag at various velocities from the induced drag curve and solving for the corresponding values of P_{Ri}, then plotting, we see an entirely different shaped curve. The induced power required curve is seen to be much flatter than the induced drag curve. See Figure 8.1(b),(a).

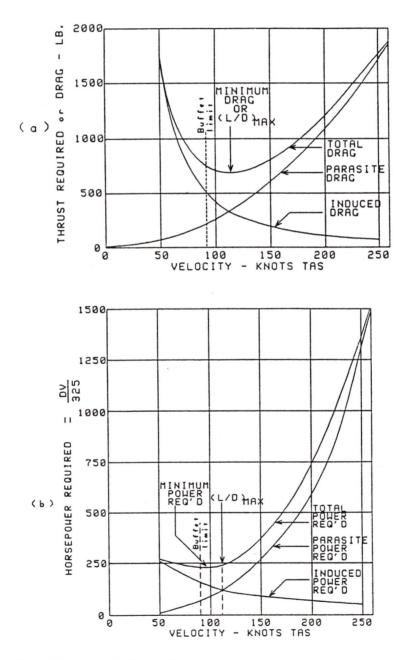

Figure 8.1   (a) Thrust-Required, (b) Power-Required curves.

The P_{Ri} curve varies with velocity as:

$$\frac{P_{Ri2}}{P_{Ri1}} = \frac{V_1}{V_2}$$
(8.1)

While the induced drag varied as the inverse of the velocity ratio "squared," the induced power required varies as the inverse of the velocity ratio "to the first power." It can be seen that there is no large increase in induced power required at low speeds as there was for induced drag.

To determine the power required to overcome the parasite drag (PRp), convert the parasite drag values at selected velocities to the corresponding power required values and plot. This curve is labeled as PARASITE POWER REQ'D in Figure 8.1b and it varies with velocity as:

$$\frac{P_{Rp2}}{P_{Rp1}} = \left(\frac{V_2}{V_1}\right)^3$$
(8.2)

The parasite drag varies directly as the "square" of the velocity ratio. The parasite power required varies directly as the "cube" of the velocity ratio. The parasite power-required curve is seen to be steeper than the parasite drag curve in Figure 8.1.

The total power-required curve is simply the addition of the P_{Ri} and the P_{Rp} curves. The P_R curve is seen to be flatter in the low speed region than the T_R curve, but it is steeper in the high speed region. This helps explain why propeller aircraft perform better than turbojets in the low speed region, but lack the high power required to fly at high speeds.

The intersection of the P_{Ri} and P_{Rp} curves still shows the $(L/D)_{MAX}$ point; however, this is no longer the lowest point on the curve. A tangent line from the origin to the curve will locate the $(L/D)_{MAX}$ point.

To re-emphasize, the lowest point on the drag curve is $(L/D)_{MAX}$. The tangent point on the power-required curve is $(L/D)_{MAX}$.

For this discussion we have chosen a "typical" propeller airplane whose power-required curve is shown in Figure 8.2.

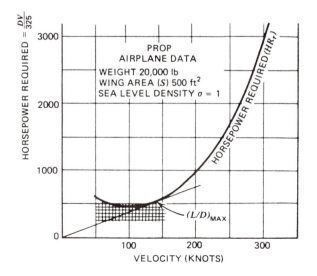

Figure 8.2   Power Required.

PRINCIPLES OF PROPULSION

The principles of propulsion are based on Newton's second and third laws. The second law is written as:

$$F = ma$$

Simply stated, an unbalanced force F, acting on a mass m, will accelerate a the mass in the direction of the force.

In propeller aircraft the force is supplied by the engine shaft which rotates the propeller. The propeller moves the air mass through the propeller disc and accelerates it.

Newton's third law states that for every action force, there is an equal and opposite reaction force. It is this reaction force that provides the thrust, T, to propel the aircraft.

The thrust was shown in equation 6.1 to be:

$$T = Q(V_2 - V_1)$$

where

T   =   thrust (lb.
Q   =   mass airflow = ρAV (slugs/sec; equation 2.6)
V_1 =   inlet (flight) velocity (fps)
V_2 =   final velocity (fps)

Figure 8.3 shows a schematic of the process of producing thrust from a propeller.

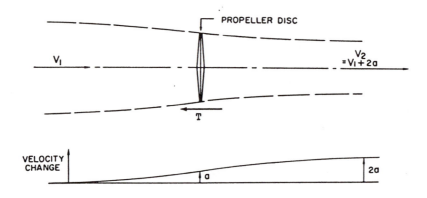

Figure 8.3   Thrust from a propeller.

Propulsion efficiency, η_P, was expressed in equation 6.2 as:

$$\eta_P = \frac{2V_1}{V_2 + V_1}$$

110

High exhaust velocity does produce high thrust, but it also produces low efficiency. High mass airflow also produces high thrust and does not reduce propulsion efficiency. Propeller aircraft process large quantities of air with only a small acceleration of the air compared with turbojets. Therefore they are much more efficient than turbojets.

POWER AVAILABLE

There are several types of horsepower produced by a power-producing engine and propeller combination:

1. *Brake horsepower*, (BHP), got its name from an early device used to measure horsepower, called a "prony brake." It is the horsepower measured at the crankshaft (piston engines) or at the turbine shaft (turbine engines).

2. *Shaft horsepower*, (SHP), is less than the BHP because of gearing losses in reducing engine RPM to propeller RPM. It is measured at the propeller shaft.

3. *Thrust horsepower*, (THP), is the usable horsepower. It is less than the SHP because of propeller efficiency loss. Do not confuse THP with thrust. Thrust horsepower is a type of horsepower and must be converted to thrust units by equation 1.6.

Turboprop aircraft produce both power and thrust. The amount of thrust produced directly by the engine is only about 15 percent of the total. They are, therefore, classified as power producers. The amount of thrust that they produce is converted into horsepower units and added to the shaft horsepower. The result is called equivalent shaft horsepower, (ESHP).

$$ESHP = SHP + \frac{TV}{325\eta} \qquad (8.3)$$

where η = propeller efficiency

Propeller efficiency, η is defined as the ratio of power output to power input:

$$\eta = \frac{THP}{SHP} \qquad (8.4)$$

Propellers can be classified into four general groups:

1. Fixed pitch. A one pitch setting blade propeller, usually of two blade design. This type of blade achieves its maximum efficiency at only one airspeed.

2. Adjustable pitch. The pitch setting can be adjusted on the ground and requires that the engine be stopped and, in some cases, that the propeller be removed from the aircraft.

3. Controllable pitch. The pitch of this type can be changed in the air by the pilot. It usually is restricted to either the takeoff setting (low pitch) or to cruise setting (high pitch). The pilot control is usually mechanical, although some models are hydraulically or electrically controlled.

4. Constant speed. This type is hydraulically or electrically controlled. The pilot selects the desired RPM with a lever and a control governor automatically changes the blade angle to maintain a constant propeller RPM. This type is the most efficient, but a maximum of about 92 percent efficiency is the best that has been obtained to date. Turboprop RPM is electronically controlled without pilot input.

In our earlier discussion of thrust available from turbojet engines (Chapter Six), we saw that the thrust output did not vary, to any great extent, with the velocity of the aircraft.

A similar relationship between brake horsepower (and shaft horsepower) and the aircraft velocity exists in power producing aircraft. However, the thrust horsepower does vary significantly with velocity. This is due to the fact that the propeller efficiency varies with velocity. Thrust horsepower is called power available, P_A.

Figure 8.4 shows a plot of power available and power required. You will note that fuel flow is also plotted on the vertical scale.

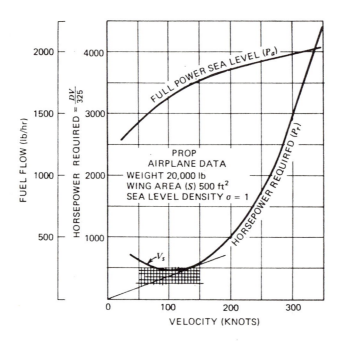

Figure 8.4   Power required and power available.

Variations of Power Available at Altitude

In discussing power available from power-producing engines, we must consider three types of powerplants:

1. turboprops,
2. unsupercharged reciprocating engines,
3. supercharged reciprocating engines.

1. In Chapter Six we discussed the effects on thrust output of turbojet engines when RPM and altitude were changed. Generally, these effects apply to any gas turbine engine (including power producers), so they will not be repeated here. In summary we can say that as altitude increases, the power available from a turboprop engine decreases and the fuel flow is also decreased.

2. Reciprocating engines do not operate at high RPM settings (except for takeoff power). The specific fuel consumption of these engines is called brake specific fuel consumption, (c_B).

c_B is the fuel flow per brake horsepower of the engine.

$$c_B = \frac{\text{fuel flow (lb./hr.)}}{\text{brake horsepower}} = \frac{FF}{BHP} \qquad (8.5)$$

Figure 8.5 shows the variation of c_B with power available. Lowest c_B values occur in the mid power (40-60 percent) range.

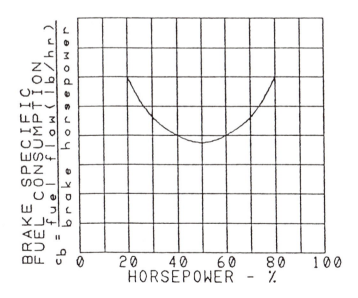

Figure 8.5   Brake specific fuel consumption vs. BHP.

High power requires high RPM and high friction. This causes high fuel flow and high c_B. Excessively low power will also cause an increase in c_B.

A very serious problem in reciprocating engines is called *detonation*. Detonation results from a sudden, unstable decomposition of fuel at high temperature and/or pressure.

Under certain combinations of these, the mixture ahead of the advancing flame front can suddenly explode instead of smoothly burning. This explosion will send strong pressure waves throughout the combustion chamber which are many times higher than normal combustion pressures. Heavy detonation can cause immediate severe structural damage to the engine.

Detonation can be detected by a rapid rise in cylinder head pressure, a drop in manifold pressure, loss of power, and loud explosive noises (similar to "ping" in car engines). Reducing throttle may actually increase power output by stopping the detonation.

3. Unsupercharged reciprocating engines are called *normally aspirated engines*. They lose power as altitude is increased and the air density is reduced. At about 19,000 feet altitude they can only develop only about one- half of the sea level horsepower.

Modern turbo superchargers use exhaust gases to rotate a centrifugal compressor. They can increase the power available so that sea level power can be obtained up to a *critical altitude* of about 19,000 feet.

ITEMS OF AIRCRAFT PERFORMANCE

Straight and Level Flight

Equilibrium conditions on an aircraft exist when the power developed by the engine is equal to the power required by the airframe for the same set of flying conditions. When these conditions are met, the aircraft will fly at a constant altitude and airspeed. Unaccelerated maximum velocity will occur at the intersection of the full power-available curve and the power-required curve. This is shown in Figure 8.4 and is at about 335 knots TAS. Of course, the assumption is made that other aircraft restrictions as to maximum velocity do not restrict flight at this airspeed.

Climb Performance

Modern propeller aircraft are not used as fighters so they do not use "zoom" techniques as do jet fighters. Therefore, steady velocity climb is assumed in this discussion. In steady velocity climb, the aircraft is in equilibrium, with all the forces along the flight path being in balance. This is shown in Figure 8.6.

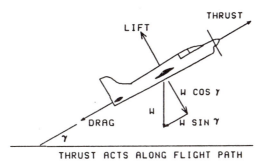

THRUST ACTS ALONG FLIGHT PATH

Figure 8.6   Forces on climbing aircraft.

The forces acting along the flight path are thrust, acting forward, drag and W sin γ, acting to the rear. W sin γ actually acts at the aircraft's CG, not as shown in the figure. For steady velocity to exist, these must be balanced.

From equation 6.3 we saw that, for equilibrium:

$$T - D - W \sin \gamma = 0$$

Rearranging we get equation 6.3a:

$$\sin \gamma = \frac{T-D}{W} = \frac{T_A - T_R}{W}$$

Angle of Climb

From equation 6.3a it can be seen that the maximum climb angle is obtained when the excess thrust is a maximum. However to avoid the confusion of using thrust curves for power-producing aircraft a different approach will be presented here.

The basic equation 1.6 for converting thrust or drag units to horsepower units can be rewritten as:

$$T = \frac{325\ HP}{V_K}$$

Adding subscripts yields:

$$T_A = \frac{325 P_A}{V_K} \quad \text{also} \quad T_R = \frac{325\ P_R}{V_K}$$

By substituting in equation 6.3a we have:

$$\sin \gamma = \frac{325\ (P_A - P_R)}{(V_K)\ (W)}$$

(8.6)

If we select values of P_A and P_R from Figure 8.4 at several arbitrary velocities, we can calculate the sine of the corresponding climb angle using equation 8.6. The climb angle itself then can be plotted against velocity as shown in Figure 8.7.

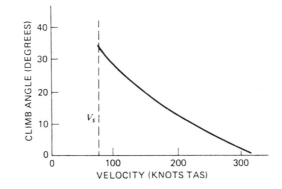

Figure 8.7   Climb angle versus velocity.

Maximum climb angle occurs at the stall speed for this aircraft. Contrast this to the thrust-producing aircraft, which has its maximum climb angle at $(L/D)_{MAX}$. The advantage of the propeller aircraft over the jet, in short field obstacle clearance is easily seen. The old saying, "Hang it on the props," is really true. A discussion of wind effect on obstacle clearance is similar to that for thrust-producing aircraft (see Figure 6.11) and so it is not repeated here.

Rate of Climb

Rate of climb is illustrated in Figure 8.8. The vertical component of the flight velocity is the rate of climb, RC.

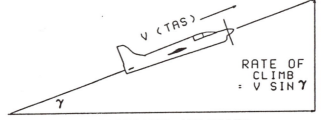

Figure 8.8 Rate of climb velocity vector.

$$RC = V \sin \gamma$$

$$RC = \frac{V_K (T-D)}{W} \quad \text{(knots)}$$

$$RC = \frac{101.3 \, (V_K) \, (T-D)}{W} \quad \text{(fpm)}$$

116

or

$$RC = \frac{101.3\ (TV-DV)}{W}$$

Substituting

$$TV = 325\ P_A \text{ and } DV = 325\ P_R$$

Therefore

$$RC = \frac{33,000\ (P_A - P_R)}{W} \tag{8.7}$$

Maximum rate of climb occurs at the velocity where maximum excess power occurs. This is shown in Figure 8.9.

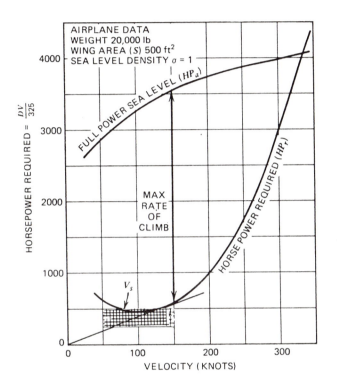

Figure 8.9   Finding maximum rate of climb.

At 150 knots the vertical distance between the P_A and the P_R curves is a maximum and thus RC is a maximum here. At this speed, $P_A = 3550$ HP, $P_R = 600$ HP, and

$$RC = \frac{33,000\ (3550-600)}{20,000} = 4867.5 \text{ fpm}$$

The climb angle here is 18.6°. Comparing this to the maximum climb angle (33.5°) speed of 75 knots, where $P_A = 3050$ HP, $P_R = 500$ HP, we find:

$$RC = \frac{33,000\ (3050-500)}{20,000} = 4207.5 \text{ fpm}$$

Endurance

To obtain maximum endurance, minimum fuel flow is required. Time in flight, not distance covered, is the objective.

$$\text{Specific endurance} = \frac{\text{fuel available (lb.)}}{\text{fuel flow (lb./hr.)}} \text{ (hours)}$$

Figure 8.10 is used to show the velocity to fly for both maximum endurance and maximum specific range for a propeller aircraft. Because fuel flow is roughly proportional to the power required, minimum P_R will produce minimum fuel flow. As mentioned earlier, this is not $(L/D)_{MAX}$ for a propeller aircraft. It is merely called minimum power required.

Mathematically it is the point where $\dfrac{(C_L)^{3/2}}{C_D}$ is maximum.

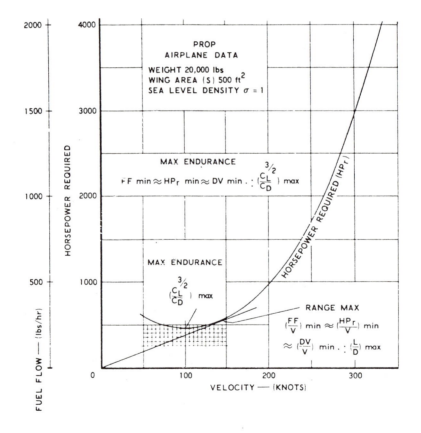

Figure 8.10 Finding maximum endurance and range.

Specific Range

To get maximum range, the aircraft must fly the maximum distance with the fuel available. The specific range must be a maximum. Specific range can be defined by the following relationship:

$$\text{Specific range} = \frac{\text{naut. mi. flown}}{\text{lb. fuel used}}$$

$$= \frac{\text{nmi/hr}}{\text{lb. fuel/hr}}$$

$$= \frac{\text{airspeed (knots)}}{\text{fuel flow (lb./hr.)}}$$

For maximum specific range:

$$(SR)_{MAX} = \left(\frac{V}{FF}\right)_{MAX} \quad \text{or} \quad (SR)_{MAX} \text{ occurs at } \left(\frac{FF}{V}\right)_{MIN}$$

The point on the P_R curve in Figure 8.10 where a straight line drawn from the origin is tangent to the curve will indicate the maximum specific range velocity. This is similar to thrust producing aircraft (Figure 6.17) with one exception. For a power-producing aircraft, this is the $(L/D)_{MAX}$ point, for a thrust-producing aircraft, it was not.

Two items of aircraft performance occur at $(L/D)_{MAX}$ for power-producing aircraft:

1. maximum engine-out glide ratio (minimum drag), and
2. maximum range.

Every pilot should realize the importance of the $(L/D)_{MAX}$. He should know the airspeed to fly, under all weight conditions, to obtain it.

Wind Effect on Specific Range

An aircraft flying into a headwind is in an unfavorable environment. Therefore the airspeed should be increased to reduce the time that the aircraft is exposed to the headwind. Conversely, an aircraft flying with a tailwind is being helped on its way and should slow down to take advantage of this favorable environment. The amount that the airspeed should be changed is shown in Figure 8.11.

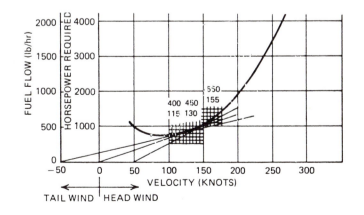

Figure 8.11   Effect of wind on range.

To find the corrected airspeed for a head wind, plot the wind on the velocity scale to the right of the origin. Draw a new tangent line from this point to the P_R curve.

Correcting for a tail wind is done in a similar manner. The tail wind is laid off to the left of the origin and a new tangent line is drawn to the curve from this point.

Total Range

The total range was explained earlier (see Figure 6.19) for thrust-producing aircraft. It is similar for power-producing aircraft.

SYMBOLS

English Symbols

BHP   Brake horsepower
c_B Brake specific fuel consumption (lb fuel/BHP)
ESHP   Equivalent SHP
P_A Power available (horsepower units)
P_R Power required
P_{Ri} Power required, induced
P_{Rp} Power required, parasite
SHP   Shaft horsepower
THP   Thrust horsepower

Greek Symbol
η (eta)   Propeller efficiency (%/100)

EQUATIONS

8.1 $\quad \dfrac{P_{Ri2}}{P_{Ri1}} = \dfrac{V_1}{V_2}$

8.2 $\quad \dfrac{P_{Rp2}}{P_{Rp1}} = \left(\dfrac{V_2}{V_1}\right)^3$

8.3 $\quad ESHP = SHP + \dfrac{TV}{325\eta}$

8.4 $\quad \eta = \dfrac{THP}{SHP}$

8.5 $\quad c_B = \dfrac{FF}{BHP}$

8.6 $\quad \sin\gamma = \dfrac{325\,(P_A - P_R)}{(V_K)\,(W)}$

8.7 $\quad RC = \dfrac{33{,}000\,(P_A - P_R)}{W}$

PROBLEMS

1. In the equation $P_R = \dfrac{T_R V}{325}$, P_R is:

    a. power required in horsepower units.
    b. power required in ft-lb/sec.
    c. resultant power (thrust-drag).

2. Power is:

    a. (force × velocity) ÷ time.
    b. work/time.
    c. (force × distance) ÷ time.
    d. both b. and c. above.

3. Power required to overcome induced drag varies:

    a. inversely with V^2.
    b. inversely with V^3.
    c. inversely with V.
    d. directly with V.

4. Power required to overcome parasite drag varies:

    a. directly with V^2.
    b. directly with V^3.
    c. directly with V.
    d. inversely with V^2.

5. Maximum rate of climb for a propeller airplane occurs:

    a. at $(L/D)_{MAX}$.
    b. at $PR_{(MIN)}$.
    c. at $C_{L(MAX)}$.
    d. at $(P_A - P_R)_{MAX}$.

6. The lowest point on the P_R curve is $(L/D)_{MAX}$.

    a. True.
    b. False.

7. Propeller aircraft are more efficient than jet aircraft because:

    a. they don't go so fast.
    b. they process more air and don't accelerate it so much.
    c. they use gasoline instead of JP fuel.
    d. V_1 is less, so propulsive efficiency is greater.

8. Turboprop aircraft are classified as power producers because:

    a. nearly all of the engine output goes to the propeller.
    b. the engine is a turbine engine.
    c. the fuel flow is proportional to the power produced.
    d. both a. and c. above.

9. Helicopters have another power requirement over fixed wing propeller airplanes. It is called:

    a. induced power required.
    b. parasite power required.
    c. profile power required.
    d. total power required.

10. Propeller aircraft get the highest angle of climb at $(L/D)_{MAX}$.

    a. True.
    b. False.

11. In Chapter Five, problem 14 you calculated the drag for a 10,000 lb. turbojet airplane for sea level standard day conditions. The values that you found are repeated below. Now calculate the values for horsepower required and plot them on graph paper.

V_K	Drag, lb	P_R
125	1233	
150	1058	
172	1020	
200	1067	
300	1720	
400	2860	

12. P_A - P_R curves for a propeller airplane are shown in Figure 8.9. Find:

    (a) V_{MAX} at full power
    (b) sine of climb angle at V_s
    (c) sine of climb angle at $(L/D)_{MAX}$
    (d) rate of climb at V_S
    (e) rate of climb at 150 knots
    (f) Velocity for maximum endurance
    (g) Velocity for maximum range
    (h) Velocity for maximum range with 50 knot headwind.

CHAPTER NINE

PROP AIRCRAFT APPLIED PERFORMANCE

The aircraft performance items that we discussed in Chapter Eight were for one weight, clean configuration, and sea level standard day conditions. In this chapter we will alter these conditions and examine the resulting performance. To simplify the explanation, only one condition will be altered at a time.

VARIATIONS IN THE POWER-REQUIRED CURVE

The basic power-required curve (Figure 8.2) was drawn for a 20,000 lb. airplane in the clean configuration, at sea level on a standard day. We will now show how the curve changes for variations in weight, configuration, and altitude.

Weight Changes

To find out how the P_R varies with weight changes, we go back to our basic relationship between P_R and T_R (drag) that we developed in equation 1.6:

$$P_R = \frac{DV}{325}$$

Adding subscripts and forming a ratio gives:

$$\frac{P_{R2}}{P_{R1}} = \frac{D_2 V_2}{D_1 V_1}$$

From equation 7.1:

$$\frac{D_2}{D_1} = \frac{W_2}{W_1}$$

and from equation 4.2:

$$\frac{V_2}{V_1} = \sqrt{\frac{W_2}{W_1}}$$

Substituting gives:

$$\frac{P_{R2}}{P_{R1}} = \left(\frac{W_2}{W_1}\right)^{3/2} \tag{9.1}$$

In addition to the increase in P_R, with an increase in weight, there also must be an increase in airspeed. This is the same as was found in equation 4.2:

$$\frac{V_2}{V_1} = \sqrt{\frac{W_2}{W_1}}$$

In plotting new P_R curves for a change in weight, it must be emphasized that both equations 9.1 and 4.2 must be applied.

Figure 9.1 shows P_R curves for 20,000 lb. and for 30,000 lb.

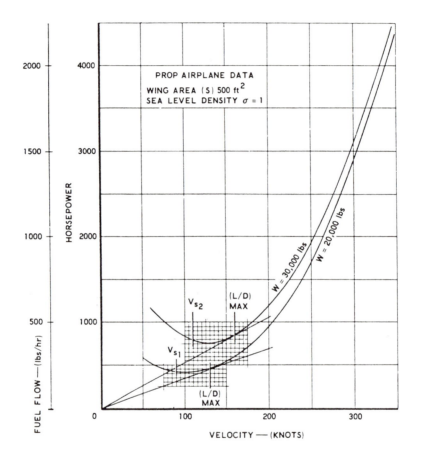

Figure 9.1   Effect of weight change on P_R curve.

Configuration Change

In going from the clean configuration to the gear and flaps down (dirty) configuration, the equivalent parasite area is greatly, increased. For this airplane it increases by 50 percent (from 9.92 to 14.9 ft²). This increases the parasite drag, and thus the parasite power required, but has little effect on the induced drag or induced power required.
There is no "short cut" method of calculating the P_R curve for the dirty condition. You must calculate the parasite drag and then the total drag for the dirty condition, and then convert this to power required by equation 1.6. The effect of configuration change is shown on the P_R curves in Figure 9.2.

124

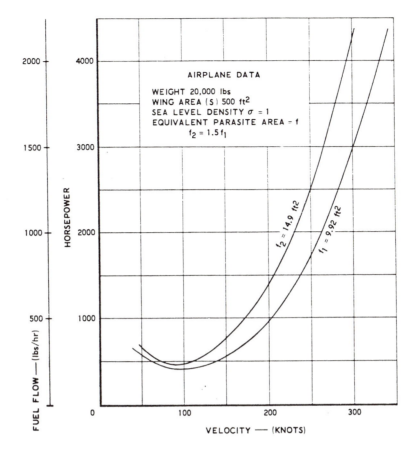

Figure 9.2 **Effect of configuration on P_R curve.**

Altitude Changes

In our discussion of the effect of altitude on the drag of an aircraft, we saw that the drag of the aircraft was unaffected by altitude, but that the true airspeed at which the drag occurred did change. Mathematically:

$$D_2 = D_1 \text{ and } V_2 = \frac{V_1}{\sqrt{\sigma_2}}$$

where subscript 1 is sea level ($\sigma_1 = 1$) and subscript 2 is at altitude. Applying the above equations to equation 1.6 and forming a ratio gives:

$$\frac{P_{R2}}{P_{R1}} = \frac{1}{\sqrt{\sigma_2}}$$

(9.2)

So, even though the drag does not change with altitude, the power required does. In addition, the velocity changes by the same amount.

$$\frac{V_2}{V_1} = \frac{1}{\sqrt{\sigma_2}}$$

(9.3)

125

In plotting new P_R curves for a change in altitude, note that both equations 9.2 and 9.3 must be applied.

All points on the sea level curve move to the right and upward, by the same amount, when correcting for altitude changes. This can be seen in Figure 9.3.

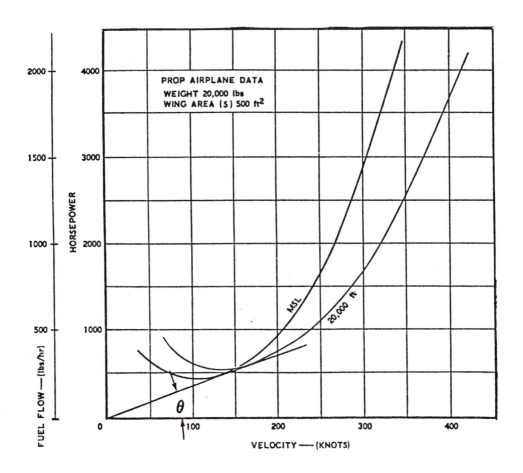

Figure 9.3   Effect of altitude on P_R curve.

Note that the line drawn from the origin, which is tangent to the curve and locates the $(L/D)_{MAX}$ point, remains tangent to the curve at all altitudes. The angle remains the same at all altitudes. The significance of this is explained later in this chapter.

Another point of interest in the altitude curves is that the curves move farther apart at the higher velocities. Each point on the sea level curve is moved to the right and also moved upward by the same multiplier, $(1/\sqrt{\sigma_2})$, so the change is greater when the velocity or power required is greater, thus points on the right are affected more than those on the left.

VARIATIONS IN AIRCRAFT PERFORMANCE

Straight and Level Flight

Weight Change. Increase in weight means an increase in P_R, but P_A is not affected, so V_{MAX} is decreased. A reduction in weight has the opposite effect and V_{MAX} will be increased.

Configuration Change. V_{MAX} in the dirty configuration is usually limited by structural strength of gear and flaps, so V_{MAX} is reduced.

Altitude Increase. The power available at altitude is affected by the type of engine and/or propeller.

If turbo supercharged reciprocating engines with constant speed props are flown below their critical altitude they can develop as much power as at sea level. In this case the V_{MAX} TAS will be increased significantly.

Unsupercharged recips with fixed props, on the other hand, will suffer a power loss above about 5,000 ft. and may experience either an increase or decrease in V_{MAX}, depending upon the intersection point of the P_A and P_R curves.

Similar reasoning can be applied to turboprop engines. The engine loses power with increased altitude, but the P_R curve moves to the right, so the intersection of the curves will determine V_{MAX}. The IAS will be lower in all cases.

Climb Performance

Angle of Climb

Weight Change. Increased weight results in increased power required while power available is not affected. Equation 8.6 shows:

$$\sin \gamma = \frac{325(P_A - P_R)}{V_K W}$$

Thus both the excess power reduction and the increased weight cause a reduction in angle of climb.

Configuration Change. Power required increases in the dirty configuration, thus angle of climb is reduced. However, the increase in P_R is much less than was the increase in T_R for the jet aircraft (see Figures 9.2 and 7.4). Again the superior performance, at low speeds, of propeller aircraft is demonstrated.

Altitude Increase. Again referring to equation 8.6, it can be seen that the increase in P_R at altitude will reduce the angle of climb, even if P_A is not reduced.

Rate of Climb

Weight Change. Rate of climb is calculated by equation 8.7:

$$RC = \frac{33,000 \, (P_A - P_R)}{W}$$

Again the increase in P_R and the increase in weight, both reduce the rate of climb at higher weights.

Configuration Change. A dirty aircraft reduces all climb performance so clean up the aircraft as soon as practical.

Altitude Increase. P_R increases so ROC decreases (equation 8.7).

Endurance

Weight Change. More power is required with increased weight. Fuel flow is proportional to P_R, so endurance is also reduced. The opposite is true when weight is decreased.

Configuration Change. Pull up your gear and flaps to improve endurance. If it is impossible to clean up the airplane, slow down to reduce parasite drag.

Altitude Increase. With a turboprop, the reduction in fuel flow at altitude will more than compensate for the increase in P_R, but if it is necessary to climb to altitude, you may use more fuel than if you stayed at a lower altitude. For a reciprocating engine the endurance at altitude is reduced.

Specific Range

Weight Change. As fuel is burned, weight is reduced, and the power required curve moves downward and to the left. This is shown in Figure 9.4. The tangent lines intersect the P_R curves at $(L/D)_{MAX}$ and indicate the maximum specific range velocity and fuel flow at this point. At a gross weight of 30,000 lb., the velocity must be 160 knots for maximum range and the fuel flow will be 425 lb/hr.
The SR here is 160/425 = 0.3765 nmi/lb.

When the weight of the aircraft has been reduced to 20,000 lb., the aircraft must fly at 130 knots to attain maximum range. The fuel flow here will be 225 lb/hr and the SR = 130/225 = 0.5778 nmi/lb. This is a 53.5 percent improvement in specific range. However, the pilot must reduce the power output of the engine and reduce the airspeed in order to achieve this range. It is only necessary for the pilot to reduce his power setting as the airspeed will automatically be reduced when this is done.

The natural tendency of a pilot is to keep the power setting the same, as fuel is burned, and to allow the velocity to increase. This is **WRONG.** This will decrease the range by an appreciable amount. For maximum range, as weight is reduced, **REDUCE AIRSPEED.**

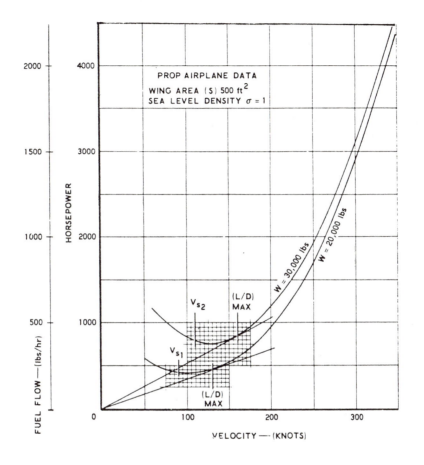

Figure 9.4   Effect of weight change on specific range.

Configuration Change. Should it be necessary to fly in the dirty configuration, slow the airplane to reduce drag.

Altitude Increase. In our earlier discussion we pointed out that an altitude increase affected both the P_R and the velocity by the same amount. We also saw that the tangent line to the curves remained the same for all altitudes. In Figure 8.10 we saw that the intersection of the tangent line and the curve indicated the maximum specific range point. Thus, as far as the airframe is concerned, the altitude has no effect on the specific range. The ratio of fuel flow to velocity is constant with altitude.

We must consider at what altitude the engine-propeller combination is most efficient. First, let's look at the unsupercharged reciprocating engine with a fixed pitch propeller. This engine can develop sea level power to about 5,000 feet. This seems to be about the altitude where maximum range is obtained. For the turbine supercharged reciprocating engine with constant speed propeller about 10,000 feet is best. The turbine engine of the turboprop aircraft likes high altitude and so it operates best at about 25,000 feet.

EQUATIONS

9.1 $\dfrac{P_{R2}}{P_{R1}} = \left(\dfrac{W_2}{W_1}\right)^{3/2}$

9.2 $\dfrac{P_{R2}}{P_{R1}} = \dfrac{1}{\sqrt{\sigma_2}}$

9.3 $\dfrac{V_2}{V_1} = \dfrac{1}{\sqrt{\sigma_2}}$

PROBLEMS

1. A lightly loaded propeller airplane will be able to glide _____ than when it is heavily loaded.

    a. farther.
    b. less far.
    c. the same distance.

2. To obtain maximum glide distance, a heavily loaded airplane must be flown at a higher airspeed than if it is lightly loaded.

    a. True.
    b. False

3. For maximum range at high altitude, a propeller driven aircraft must be flown at _____ true airspeed than at sea level.

    a. less.
    b. more.
    c. the same.

4. Referring to Figure 9.3, the specific range for a propeller airplane is _____ at altitude as at sea level.

    a. less.
    b. more.
    c. the same.

5. The pilot of a propeller airplane is flying at the speed for best range under no wind conditions. A head wind is encountered. To obtain best range the airplane must now:

    a. speed up by the amount of the wind speed.
    b. slow down by the amount of the wind speed.
    c. speed up by less than the wind speed.
    d. slow down by more than the wind speed.

6. As a propeller airplane burns up fuel, to fly for maximum range, the airspeed must:

    a. remain the same.
    b. be allowed to speed up.
    c. be slowed down.

7. The turboprop aircraft has its lowest specific fuel consumption at about 21,000 feet altitude because:

    a. the turbine engine wants to fly at high altitudes.
    b. the propeller has higher efficiency at lower altitudes.
    c. this is a compromise between a. and b. above.

8. The Power required curves for an increase in altitude show that the:

    a. P_R remains the same as altitude increases.
    b. P_R increases by the same amount as the velocity.
    c. PR increases but the velocity does not.

9. A propeller aircraft in the dirty condition shows that the P_R moves up and to the left over the clean configuration. This is because:

    a. the increase in the induced P_R is more at low speed.
    b. the increase is mostly due to parasite P_R.
    c. they both increase by the same amount.

10. The increase in the P_R curves for a weight increase is greater at low speeds than at high speeds because the increase in:

    a. induced P_R is greatest.
    b. parasite P_R is greatest.
    c. profile P_R is greatest.

11. Using Figure 9.1 find the velocity for best range for the airplane at 30,000 lb. and 20,000 lb. G.W.

12. Using Figure 9.1 calculate the specific range for the airplane at both weights if it is flying at the best range airspeed.

13. Using Figure 9.1 calculate the specific range for the 20,000 lb. airplane if the throttle is not retarded from the 30,000 lb. best range position and the airspeed is allowed to increase as fuel is burned and weight is reduced. Compare your answer to that of problem 12 and state your conclusions.

14. Using Figure 9.3 calculate the specific range for this airplane if it is flying at best range airspeed at mean sea level (MSL) and also at 20,000 ft. altitude.

CHAPTER TEN

HELICOPTER AERODYNAMICS

INTRODUCTION

A helicopter looks strange as compared to a conventional fixed-wing aircraft, however the same principles of flight apply to both. In Chapter Three we learned that Leonardo da Vinci stated that air passing around an airfoil produces the same aerodynamic forces if the airfoil is moving through still air or whether the air is moving past a stationary air foil.

We can carry this principle further in saying, "an airfoil will produce the same aerodynamic force whether the airfoil is being moved through the air by moving the entire fixed wing aircraft, or by moving the airfoil while the rotary wing aircraft remains stationary."

MOMENTUM THEORY OF LIFT

Bernoulli's equation of lift is still a valid theory of how a helicopter produces lift, however because the velocity of each rotor blade segment varies as the distance from the rotor hub, it is simpler to discuss the lift by using the momentum theory.

The momentum theory is based upon Newton's laws. The second of these states that an unbalanced force acting upon a body produces an acceleration of the body in the direction of the force. The "body", in the case of the flow of air through a helicopter rotor system, is the mass of air. The unbalanced force is produced by the rotor blades being driven by the engine. The result is an acceleration of the air down through the rotor disk.

Newton's third law stated that, "for every action force there is an equal and opposite reaction force." Thus by causing the air to be forced downward, lift was developed upward. This is quite similar to the production of thrust from an aircraft propeller or jet engine.

Figure 10.1 shows a schematic of this theory. In the figure:

V_o = Original velocity at about two rotor diameters above the rotor.

V_i = Induced velocity at the rotor.

V_f = Final velocity at about two rotor diameters below the rotor.

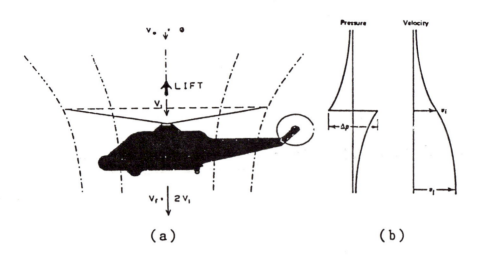

Figure 10.1 Momentum theory airflow: (a) schematic, (b) pressure and velocity distribution.

It can be seen from the figure that there is lower pressure above the rotor disk and higher pressure below the disk, with a sudden increase at the disk. The velocity increases gradually from approximately zero at about two rotor diameters above the disk to a maximum at about the same distance below the disk.

AIRFOIL SELECTION

It was shown in Chapter Three that cambered airfoils develop a nose down pitching moment when subjected to an airflow while symmetrical airfoils do not (see Figs. 3.9 and 3.10). These pitching moments change in value when the angle of attack changes. Later we will discuss the fact that during the rotation of the blades the angle of attack is constantly changing.

Rotor blades generally lack the stiffness to resist this moment and will twist if they are cambered. Twisting of the blades causes changes in AOA and lift which complicates the problem, so it is avoided by using symmetrical airfoil sections. Another advantage of the symmetrical airfoil is that both the center of pressure and the aerodynamic center are at the same location and do not move with AOA changes. So most rotor blades have symmetrical airfoil sections. Figure 10.2 shows a typical helicopter rotor airfoil.

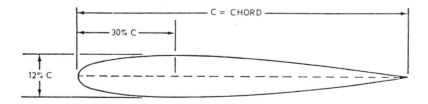

Figure 10.2 NACA 0012 airfoil.

In order to reduce pitching moments on rotor blades caused by such factors as center of gravity location, and axis of feathering (axis of rotation) location, it is desirable that they coincide with the center of pressure/aerodynamic center location. This is illustrated in Figure 10.3.

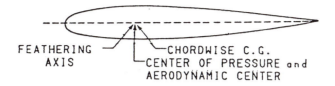

Figure 10.3   Location of critical forces on airfoil.

FORCES ON ROTOR SYSTEM

When the blades of a helicopter are not turning there is a noticeable droop to them. They do not have the inherent strength of fixed wings to prevent this droop. When the blades are rotated, however, centrifugal force causes them to straighten as is shown schematically in Figure 10.4. The force on only one blade is shown for simplicity.

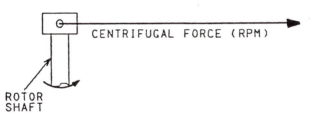

Figure 10.4   Centrifugal force straightens rotor blade.

To produce the lift required to fly the helicopter a vertical vector must be exerted upon the rotor shaft. This is shown in Figure 10.5

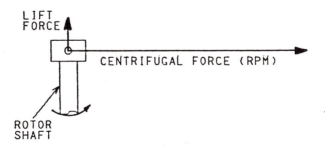

Figure 10.5   Lift force and centrifugal force.

In order to produce the vertical lift vector shown in Figure 10.5 the angle of attack of the symmetrical rotor blades must be set at a positive angle of attack. The blades are now being acted upon by two force vectors. (1) The centrifugal force vector and (2) the lift vector. The blades must assume the position of the resultant of these two vectors as shown in Figure 10.6. The lift vector has been exaggerated in the figure for clarity.

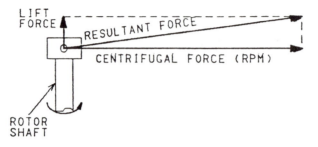

Figure 10.6   Resultant of lift and centrifugal forces.

The centrifugal force is much larger than the lift force. In fact, the lift force is only about 7% of the centrifugal force. To illustrate this, assume that the helicopter weighs 7,000 pounds and has two blades. The lift force on each blade would be 3,500 pounds while the centrifugal force would be about 50,000 pounds. This is shown in Figure 10.7.

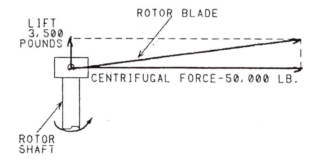

Figure 10.7   Forces acting on lifting blade.

Assuming that the rotor blades are hinged so as to allow upward rotation, the entire lifting rotor system is shown in Figure 10.8.

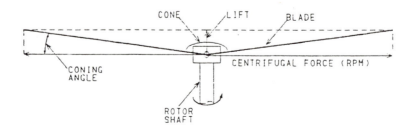

Figure 10.8   Entire lifting rotor system.

136

Of particular interest is the *coning angle* shown between the two blades and the horizontal. When viewed from the side the rotating blades form a visual disk. A line drawn between the blade tips is called the *tip path plane*. Figure 10.9 shows a hovering helicopter at a light weight.

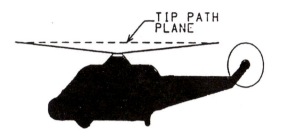

Figure 10.9   Hovering helicopter at light weight.

Under light weight conditions the coning angle is small because the lift force is relatively small as compared to the centrifugal force.

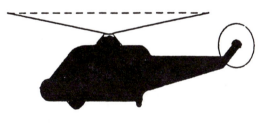

Figure 10.10   Hovering helicopter at heavy weight.

In Figure 10.10 the heavy weight requires more lift forces and, if the RPM is the same, the centrifugal forces remain the same as in the light weight condition, thus the coning angle is much larger.

THRUST DEVELOPMENT

Up to this point we have been discussing the forces on the rotor system while the helicopter was in a hover. To move the aircraft in any horizontal direction we must tilt the entire rotor disk in the desired direction. We have been calling the vertical force, *lift*. Now we will see that in addition to the force required to overcome the weight, we also obtain a component force which we call directional thrust. Therefore we rename the lift force. We call it *total thrust*.

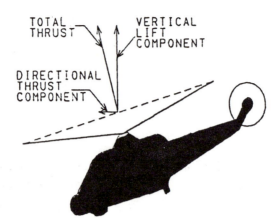

Figure 10.11   Forward flight forces.

Figure 10.11 shows the total thrust vector resolved into the vertical lift component and the forward thrust component.

One of the first reactions that results from translating from a hover to flight in any direction is a loss of altitude.

This is due to the fact that the total thrust is no longer acting opposite to the weight of the helicopter as it was in the hover. Consequently the effective lift is reduced and, if corrective action is not taken, the helicopter will hit the ground. This is shown in Figure 10.12.

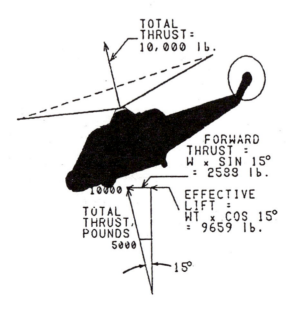

Figure 10.12   Lift component of 10,000 lb. total thrust at 15°.

The effective lift of 9659 pounds will not support the weight of 10,000 pounds so the helicopter will lose altitude. The pilot must increase total thrust to 10,353 pounds to have the effective lift equal the weight (10,353 cos 15° = 10,000).

138

HOVERING FLIGHT

Hovering means maintaining a constant position over the ground, usually at a height of a few feet above the ground. To hover, the helicopter must supply lift equal to the weight of the helicopter. To accomplish this the blades must be rotating at high RPM and the angle of attack of all of the blades is increased by increasing the angle of attack with the collective stick. We will discuss the control devices more in detail later.

The lift and weight forces are equal and opposite in a no-wind hover condition and the tip-path plane remains horizontal. If the angle of attack is increased by the pilot lifting the collective the helicopter will climb vertically. This also requires more power or the RPM will decrease.

In some helicopters there is an interlink between the collective and the throttle which automatically increases the throttle when the collective is raised. In others, the pilot must add throttle when collective is raised.

In fixed-wing aircraft the speed of the airflow over the wings is determined primarily by the airspeed of the airplane. This is not so with rotary-winged aircraft. The speed of the airflow over the rotors is determined not only by the airspeed of the helicopter, but also by the speed of rotation of the rotor blades. Let us first examine the case where the helicopter is in a hover and thus the airspeed is zero.

Hovering Blade Velocity

The velocity of each blade section depends upon its location with respect to the rotor shaft. The local velocity is a function of the distance from the shaft and the RPM of the shaft.

Figure 10.13 shows the velocity vectors of a typical two bladed rotor system having twenty foot rotors and operating at about 322 RPM. The rotor tip speed is about 400 knots.

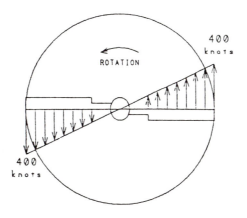

Figure 10.13   Rotor velocity distribution in hover.

Blade Twist

The relative wind of each blade is equal and opposite to the blade velocity. The speed of the blade increases as its distance from the shaft and, if each section operated at the same AOA, the lift would increase as the speed squared. (Remember the lift equation?) The blades are designed with a twist to prevent this uneven lift distribution. They have higher AOA at the root, gradually decreasing toward the tip.

Figure 10.14 shows the lift distribution on an untwisted blade and on an ideally twisted blade. Note that the twisted blade develops more lift near the root and less lift at the tip than the untwisted blade. The linear increase in lift of the twisted blade is desired.

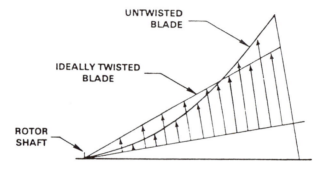

Figure 10.14   Lift distribution on twisted/untwisted blade.

Ground Effect

In the early days of helicopter development it was found that it took much less power to hover a helicopter close to the ground than at a distance above the ground. Early theory attributed this to a build up of a high pressure "bubble" beneath the helicopter upon which the helicopter balanced. This also seemed to explain why the helicopter lost altitude as it moved in any direction. It "slid off the bubble". We have already disclaimed this theory as to the loss of lift (see Fig.10.12), now we will explain the reduction of power required to hover in *ground effect*.

In Chapter Five we discussed induced drag and ground effect for fixed-wing aircraft. In case you missed it, please review those sections of the chapter. They apply to rotary-wing aircraft just as much as they do to fixed-wing aircraft.

Figure 10.15 shows a helicopter hovering out of ground effect and Figure 10.16 shows one hovering in ground effect.

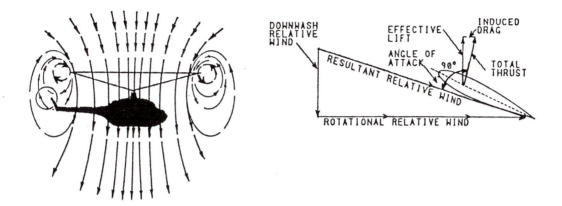

Figure 10.15   Hovering out of ground effect.

Rotor tip vortices form when the rotors are producing lift. This is quite similar to the wing tip vortices discussed in Chapter Five (see Figure 5.5). The vortices cause the air to be accelerated downward causing a downwash behind the wing or rotor. In Figure 10.15 the downwash is shown and is labeled as *DOWNWASH RELATIVE WIND*. This vector affects the *ROTATIONAL RELATIVE WIND* and the *RESULTANT RELATIVE WIND* is as shown in the figure. The *TOTAL THRUST* vector is 90° to the resultant relative wind and is tilted backward from the vertical. The *EFFECTIVE LIFT* is vertical and has a lower value than the total thrust. The rearward *INDUCED DRAG* vector is also large. To summarize, Downwash results in loss of effective lift and an increase in induced drag.

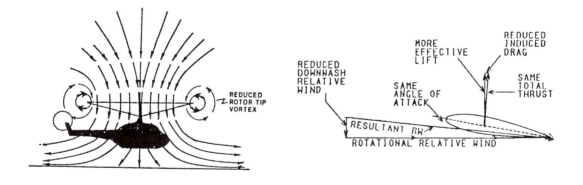

Figure 10.16   Hovering in ground effect.

If the helicopter is hovered close to the ground (or water) the downwash is physically reduced as shown in the above figure. The downwash relative wind is reduced and the angle that the resultant relative wind makes with the rotational relative wind is also reduced. Thus the effective lift is increased and the induced drag is reduced.

Torque

In American helicopters the rotor blades rotate in a counter-clockwise direction, as viewed from above. In accordance to Newton's third law of action and reaction, the fuselage tends to rotate in the opposite direction. This is called torque and must be counteracted before flight is possible. There are several ways of doing this. One method, is to use a dual rotor system. These can be either tandem or side-by-side intermeshing rotors which rotate in opposite directions. Another method is to power the rotors from the rotor tips rather than from the central shaft. This requires some sort of pulse-jet or ram-jet propulsion system. These have proven successful in experimental helicopters but are very noisy. The most common form of canceling the torque effects is the anti-torque rotor shown in Figure 10.17.

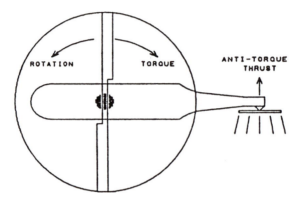

Figure 10.17   Anti-torque rotor.

It can be seen from this figure that the helicopter is not in equilibrium. There is a balance of moments, but not a balance of forces. The sideward force of the anti-torque rotor will cause the helicopter to drift to the right unless it is balanced by a force to the left. Figure 10.18 shows a method of doing this.

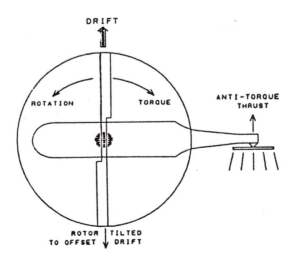

Figure 10.18   Correction for anti-torque rotor drift.

ROTOR SYSTEM CONSTRUCTION

Before we discuss forward flight, we will examine the three different rotor systems.

Rigid Rotor

The simplest system is the rigid rotor. It has but one degree of freedom. The angle of attack of the blades can be changed or *feathered*, but no other movement is allowed.

Figure 10.19 is a schematic of the rigid rotor. If a blade is required to move up or down (flap) or to move forward or backward (hunt), these movements require that the blade be bent (flexed).

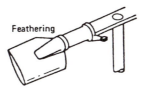

Figure 10.19   Rigid rotor system.

Semi-Rigid Rotor (See-Saw)

The semi-rigid rotor system has two degrees of freedom. In addition to the feathering axis there is also freedom of flapping. No provision for hunting is available.
Figure 10.20 shows a schematic of the semi-rigid system.

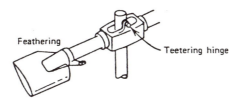

Figure 10.20   Semi-Rigid rotor system.

Fully Articulated Rotor

Three degrees of freedom are provided by the articulated rotor system. The vertical hinges allow hunting, horizontal hinges allow flapping, and rotation of the blades allow them to change angle of attack. Figure 10.21 shows this system.

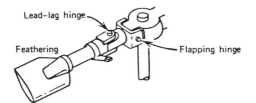

Figure 10.21   Articulated rotor system.

FORWARD FLIGHT

Early experiments in helicopter development succeeded in producing an aircraft that could hover, but failed to achieve forward flight. The helicopter would roll over when it moved from a hover in any direction. To understand what was causing this we must investigate *Dissymmetry of Lift*.

Dissymmetry of Lift

The airflow at the rotor tips that was caused by the rotation in a hover is shown in Figure 10.22.

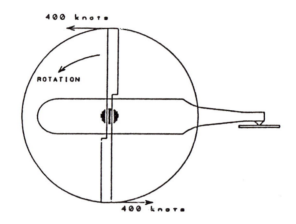

Figure 10.22   Rotor tip velocities in a hover.

As the helicopter starts to move in a horizontal direction the airflow over the rotors is changed. For simplicity, consider that the helicopter is moving forward at 100 knots.

The blade that is moving toward the front of the helicopter is called the *advancing blade* and the blade that is moving toward the rear is called the *retreating blade*.

The advancing blade tip has an airspeed of 500 knots, shown in Figure 10.23, while the retreating blade tip speed is only 300 knots.

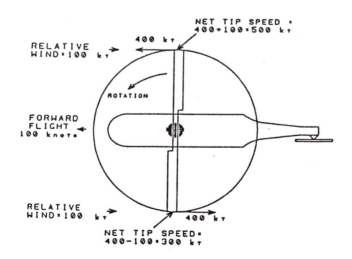

Figure 10.23   Blade tip velocity in forward flight.

144

If both blades are operating at the same AOA, the advancing blade will develop more lift than the retreating blade. This condition is *dissymmetry of lift* and cannot be tolerated.It is quite similar to flying a fixed wing airplane which has the ailerons jammed in the full left wing down/right wing up position.

The rolling moment that is produced is shown in Figure 10.24.

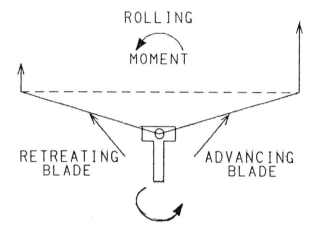

Figure 10.24   Rigid rotor rolling moment in forward flight.

Blade Flapping

The first successful solution to this problem was made by Juan de la Cierva in his experimental autogyros from 1923 to 1935. He adapted the principle of designing his rotors so that they could pivot up or down as they rotated. This motion is called *blade flapping*.

As a blade leaves the tail position and advances around the right side of the helicopter, it is influenced by the increasing airspeed and produces more lift and climbs (flaps) upward. This climbing action changes the relative wind and reduces the angle of attack and thus the lift. In a like manner the retreating blade, as it moves from the nose of the helicopter toward the tail, experiences a reduction in airspeed, loses lift and flaps downward. This results in an increase in angle of attack and an increase in lift. The final result is that the lift is equalized and the rolling moment never materializes and dissymmetry of lift does not exist.

Both the semi-rigid (see-saw) and fully articulated rotor systems have flapping hinges which automatically react to the changes in relative airspeed during blade rotation.

Figure 10.25(a) shows the angle of attack changes on the advancing blade and Figure 10.25(b) shows these changes on the retreating blade.

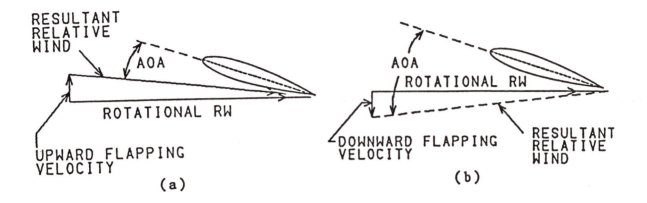

Figure 10.25   Angle of attack and flight path changes: (a) advancing blade, (b) retreating blade.

Figure 10.25(a) shows that the increasing speed of the advancing blade causes the blade to flap up. As relative wind is always equal and opposite to the flight path, the effect of this upward movement is to cause the resultant relative wind to be inclined downward. This effectively reduces the angle of attack of the advancing blade.

Figure 10.25(b) shows that the decreasing speed of the retreating blade causes a loss of lift and the blade flaps down. The effect of this downward movement is to cause the resultant relative wind to be inclined upward. This effectively increases the angle of attack of the retreating blade.

High Forward Speed Problems

As the forward speed of the helicopter is increased the relative wind speed over the advancing blade tip increases, while the relative wind speed over the retreating blade tip become less. One of two possible problems arise at some critical forward speed. Either (1) the advancing blade tip approaches supersonic speed or (2) the retreating blade approaches stall speed. We will now investigate these possibilities further.

Advancing Blade Compressibility

If the airspeed over an airfoil exceeds a certain value, called the critical Mach number, shock waves form on the upper surface of the airfoil. These shock waves cause the air to separate from the airfoil and a high speed stall results. This causes a loss of lift and an increase in drag. A more detailed discussion of high speed problems is found in Chapter Seventeen.

Retreating Blade Stall

As the helicopter's forward speed increases the relative wind over the retreating blade decreases. The resulting loss in lift causes the blade to flap down farther and the effective angle of attack increases. At some high angle of attack the blade begins to become stalled. The tip stalls first and, if no corrective action is taken, the stall progresses inward. When approximately 25% of the disk is stalled control of the helicopter is lost. Figure 10.26 shows the stall region and some typical angles of attack which the blades attain as they rotate.

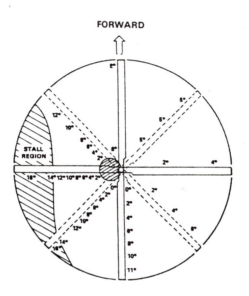

Figure 10.26   AOA distribution during retreating blade stall.

Both advancing blade compressibility and retreating blade occur under similar conditions:

* High forward speed.
* Heavy gross weight.
* Turbulent air.
* High density altitude.
* Steep or abrupt turns.

Both give warning by exhibiting abnormal vibrations under one or more of the above conditions.

They differ as to rotor RPM:
Advancing blade compressibility occurs at high rotor RPM.
Retreating blade stall occurs at low rotor RPM.

If corrective action is not taken:
Advancing blade compressibility will pitch nose down.
Retreating blade stall will pitch nose up and roll the helicopter to the left.

Corrective action for both Advancing Blade Compressibility and Retreating Blade Stall consists of:

* Checking and adjusting rotor RPM.
* Slowing the helicopter by reducing power, **not by flaring the helicopter.**
* Lower blade AOA by using down collective.
* Reduce bank angle, if in a turn.
* Reduce "G" with down collective.

Gyroscopic Precession

The rigid rotor helicopter acts more like a true gyroscope than either the see-saw or fully articulated types. The flapping hinges make the difference. However, all three types do exhibit some of the characteristics of a gyroscope as shown in Figure 10.27.

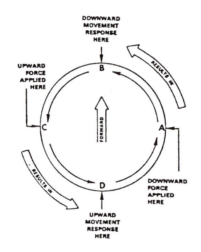

Figure 10.27 Gyroscopic precession.

A force applied to a rotating disk causes the disk to react 90 degrees later in the direction of rotation. This explains the pitch-up and pitch-down reactions to advancing blade compressibility and retreating blade stall mentioned above.

In the case of either of these the blade stalls and lift is lost. It is easy to see why the retreating blade stalls at high angle of attack, so we will discuss that condition first. The blade is at the nine o'clock position when maximum angle of attack is encountered (see Fig.10.26). Gyroscopic precession (Fig.10.27) shows the reaction occurs 90° later at the six o'clock or tail position causing a down load at the tail resulting in pitch-up.

Although it may seem strange that low angle of attack can produce stall, that is exactly what happens when critical Mach number is exceeded. Lift is lost at the three o'clock position and the reaction occurs at the twelve o'clock position causing pitch-down.

Blade Lead And Lag

Fully articulated rotor systems have vertical hinges that allow the blades to move forward and backward. This motion is called *hunting*. The hinges allow this action without requiring the blades to develop large bending stresses. The rigid and semi-rigid systems do not have these hinges.

When the helicopter is moving horizontally, the blade pitch angles are constantly changing as the blades rotate to eliminate the dissymmetry of lift. As the pitch changes, so does the drag on the blades which causes the hunting action.

There is another force called *Coriolis force* which also causes the blades to lead and lag. There is a law of physics called the law of conservation of angular momentum. The law states that the motion of a rotating body will continue to rotate with the same rotational velocity unless acted upon by some external force. If the mass is moved farther from the center of rotation it will decelerate. If the mass is moved closer to the center of rotation it will accelerate.

As a rotor blade flaps up, its center of gravity moves closer to the drive shaft and it will speed up. When the blade flaps down, the opposite occurs. Figure 10.28 shows this movement of the CG. The right rotor blade of the bottom sketch has flapped up and distance d is less than b. The blade will speed up (lead). The left blade has flapped down and the distance c is greater than a. The blade will slow down (lag). Figure 10.29 shows hunting about vertical hinge.

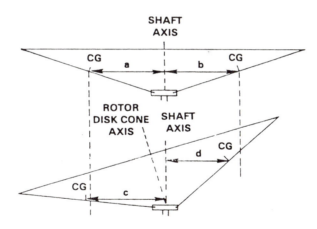

Figure 10.28   CG radius change with flapping motion.

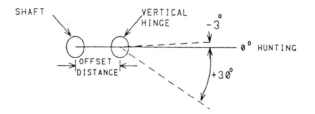

Figure 10.29   Hunting motion of fully articulated blade.

149

HELICOPTER CONTROL

The control of a helicopter consists of movement of the aircraft along the three principle axes and rotation about these axes. The rotation movements are called *pitch, yaw* and *roll.* These are discussed in some detail in Chapters Fifteen and Sixteen. In the helicopter the movement along the longitudinal axis (forward and rearward flight) and along the lateral axis (sideward flight) is controlled by the *cyclic* stick. Movement along the vertical axis (vertical flight) is controlled by the *collective* stick. Pitch and roll are also controlled by the cyclic, while yaw is controlled by the foot pedals. This book is not intended to be a flight manual, so we will describe the action of these controls, rather than a description of how to use them.

Rotor Head Control

Rotor control can be accomplished by direct tilting of the hub or by changing blade pitch. Blade pitch changing can be done by the swash plate, aerodynamic servo tabs, auxiliary rotors, jet flaps, and pitch links attached to a control gyro. The most common method is the *swash plate.* The swash plate is shown schematically in Figure 10.30.

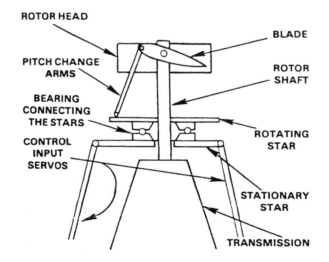

Figure 10.30   Swash plate schematic.

The stationary star and the rotating star are connected together by the bearing. This allows the rotating star, pitch change arms, blades, and rotor head to be rotated by the rotor head. The stationary star, control input servos, and transmission do not rotate. Both the rotating star and stationary star can move up and down on the rotor shaft. They both can also be tilted at various angles, but they always remain parallel to each other.

Vertical movement of the swash plate is controlled by the collective stick. If the pilot raises the collective, the pitch angle of all blades is increased and the helicopter climbs vertically. Down collective has the opposite effect. The cyclic stick tilts the stationary star through the control input servos. The rotating star remains parallel to the stationary star and changes the pitch angle of each blade by a different amount. This causes flapping of the blades and results in forces which change the tip path plane as shown in Figure 10.31. Forward cyclic movement is shown.

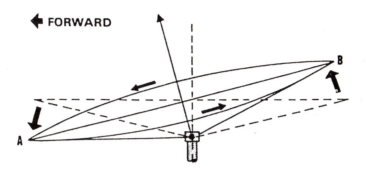

Figure 10.31   Rotor flapping caused by cyclic stick movement.

Both cyclic and collective inputs can be combined so that horizontal and vertical motions can be coupled.

Power is controlled by a motorcycle type twisting hand grip on the collective stick.

Directional Control

Foot pedals change the blade angles of the anti-torque rotor and changing its trust output and thus changing the heading of the helicopter. Figure 10.32 shows that the tail rotor is subject to dissymmetry of lift, similar to the main rotor system. A see-saw rotor hub allows blade flapping to eliminate this.

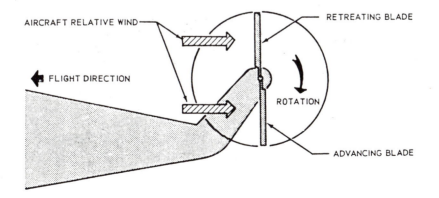

Figure 10.32   Tail rotor dissymmetry of lift.

HELICOPTER POWER-REQUIRED CURVES

In the case of fixed wing aircraft, we have seen that the power required consists of the power required to overcome induced drag and the power required to overcome the parasite drag. Rotary wing aircraft have another kind of drag. this is the drag caused by the rotation of the rotor system. The power required to turn the rotors is given a new name, *profile power required*, (P_{RO}). It is the power required to overcome the profile drag of the rotor system. Unlike the parasite drag of the helicopter, it occurs without any forward motion of the aircraft. The total power required is shown in Figure 10.32. It is the sum of the power required to overcome the induced drag, parasite drag, and profile drag.

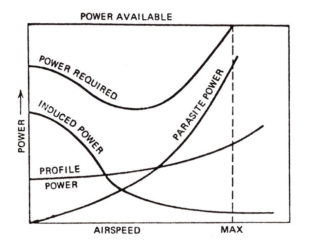

Figure 10.32   Helicopter power available and power required.

Power available has several small variations with helicopter speed, but for the sake of this discussion it is assumed to be constant.

One other difference between rotary and fixed wing aircraft is the ability of the helicopter to fly at zero airspeed. The power required curves do not terminate at stall speed, but continue down to zero airspeed.

Translational Lift

It can be seen from Figure 10.32 that there is a noticeable reduction in the total power required curve as airspeed increases from zero (hover). This is caused by *translational lift*. The efficiency of the hovering rotor system is improved as the helicopter moves forward. In the hover there is much turbulence caused by the tip vortices. As the aircraft moves into a region of undisturbed air and the vortices are left behind. The airflow becomes more horizontal and the efficiency of the rotor system is improved.

SETTLING WITH POWER

One of the most dangerous flight conditions that the helicopter pilot can encounter is known as *settling with power* or more technically known as the *vortex ring state*.

In this condition the helicopter is descending into its own downwash. This can happen when the aircraft is making a vertical or near vertical descent with low forward speed. The rotor system is under some power from the engine, not in an autorotation.

Assume that a helicopter pilot is attempting to land at a site surrounded by high trees or other obstructions. He can't make a normal approach maintaining some forward speed so he comes to a hover at some height above the ground. The airflow through the rotor system is as shown in Figure 10.33.

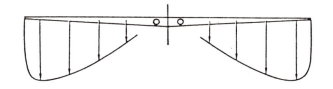

Figure 10.33   Induced flow velocity in a hover.

The maximum downward velocity is at the blade tips where the blade airspeed is the highest and decreases nearer the rotor shaft. As the helicopter descends it is acted upon by an upward relative wind which counteracts the induced flow. The resulting induced flow velocity may be as shown in Figure 10.34.

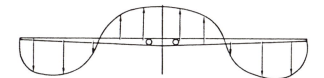

Figure 10.34   Induced flow velocity during vortex ring state.

It is obvious that, with upward and downward air flows in opposite directions, there is no lift on the helicopter and it is a free falling body. Corrective action consists of increasing power (if available) to re-establish down flow. An alternate solution is to lower collective and increase forward speed. The normal tendency to increase collective pitch while applying power is **wrong.** This action can aggravate the power settling.

If it is absolutely necessary to make a landing in a site where a forward speed of at least 15 knots can not be maintained and there is sufficient power available, the aircraft could be brought to a hover and a very low rate of descent be used.

AUTOROTATION

During forward flight the airflow through the rotor system and the force vectors are as shown in Figure 10.35.

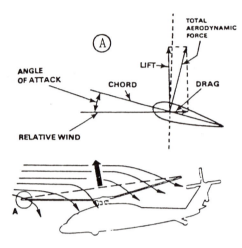

Figure 10.35   Airflow and force vectors in forward flight.

To maintain constant rotor RPM the drag vector must be counteracted by an equal and opposite force supplied by the engine. If the engine fails or is deliberately disengaged from the rotor system, some other force must be applied to sustain the rotor RPM so that the helicopter can continue flying until it reaches the ground. This unpowered flight is similar to the glide of an fixed-wing aircraft and is called *autorotation*.

The force that keeps the blades rotating is generated by lowering the collective pitch and thus reducing the angle of attack of all blades. As the helicopter loses altitude, the airflow through the rotor disk changes as seen in Figure 10.36.

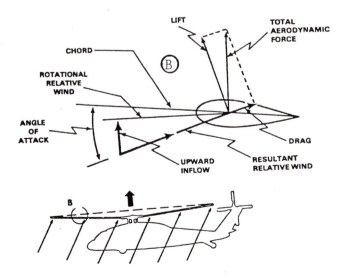

Figure 10.36   Airflow and forces in steady state descent.

The airflow during descent provides the energy to overcome the blade drag and turn the rotor. The helicopter is trading the potential energy that it has due to its altitude into kinetic energy used to turn the blades. Later we will see how this kinetic energy is used to cushion the landing.

Note that the total aerodynamic force is now vertical. The forward component of the lift vector equals the backward component of the drag vector. There are no unbalanced forces in the horizontal direction and RPM will remain constant.

To make a successful landing, the pilot must stop the forward airspeed and reduce the rate of descent just before landing. Both of these actions can be started by flaring the aircraft. This is done by moving the cyclic stick to the rear. This tilts the rotor disk to the rear and slows the forward speed. This is shown in Figure 10.37.

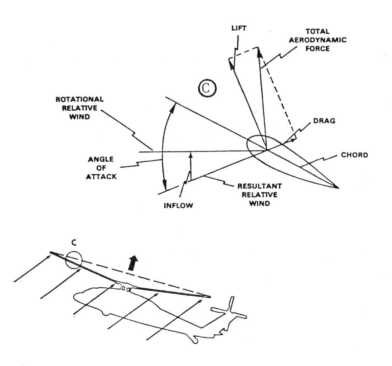

Figure 10.37   Airflow and forces during autorotative flare.

There is also an increase in rotor RPM which will help to cushion the landing. After forward speed stops, the cyclic is returned to neutral and the final vertical velocity is further reduced by using up collective.

Although vertical autorotations can be made, they are seldom attempted. The pilot has much better visibility of the landing zone and better directional control of the helicopter if forward velocity is maintained.

PROBLEMS

1. Most helicopter rotors have symmetrical airfoils because:

    a. they produce more lift than cambered airfoils.
    b. they can produce both upward and downward lift.
    c. they do not develop pitching moments.
    d. all of the above.

2. The lift theory most often used in helicopter aerodynamics is:

    a. the gyroscopic precession theory.
    b. the momentum theory.
    c. Bernoulli's theory.
    d. settling with power theory.

3. A helicopter can hover near the ground with less power than it can hover away from the ground because:

    a. the air is denser near the ground.
    b. a high pressure air "bubble" is produced below it.
    c. the rotor blades have less induced drag.
    d. none of the above.

4. The primary stress in rotor blades in flight is:

    a. tension.
    b. bending.
    c. torsion.
    d. shear.

5. As a helicopter moves forward from a hover to a speed of about 15 knots, the power required is reduced because:

    a. the efficiency of the rotor is improved.
    b. it is moving into a region of undisturbed air.
    c. the tip vortices are left behind.
    d. all of the above.

6. If a helicopter is flying at high speed and the engine fails the pilot should enter an autorotation by slowing the helicopter by using back cyclic then reducing blade AOA with the collective.

    a. True
    b. False

CHAPTER ELEVEN

HAZARDS OF LOW SPEED FLIGHT

Throughout the history of aviation, the slow speed region of flight has been the most hazardous. The reasons for this are quite different for straight winged aircraft and for swept winged aircraft. Examination of the coefficient of lift curves for each type of aircraft will help explain some of these differences.

Figure 11.1 shows the curves for a straight wing, propeller driven aircraft and a swept wing, turbojet aircraft.

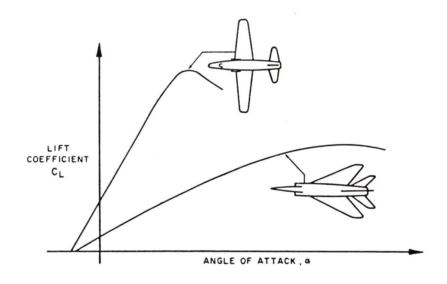

Figure 11.1 **Effect of sweepback on C_L - α curves.**

Four differences can be seen from these curves.

1. The straight winged aircraft has a higher value of C_L. Recall the basic lift equation 4.1:

$$L = \frac{C_L \, \sigma V^2 S}{295}$$

It can be seen that if the lift equals the aircraft weight, stall speed, V_S, will be a minimum when the value of the lift coefficient is a maximum, $C_{L(MAX)}$. The higher the value of $C_{L(MAX)}$, the lower the stall speed will be. This means that the straight winged aircraft has a lower stall speed, for the same weight and wing area.

2. The swept winged aircraft must fly at a higher AOA to achieve maximum lift.

3. There is a sudden reduction of C_L for the straight winged aircraft at stall, but not for the swept winged aircraft.

4. The straight winged aircraft is more sensitive to AOA change.

STALL PATTERNS

The lift distribution in the spanwise direction is shown in Figure 11.2. It is in the shape of an ellipse. The lift is zero at the wing tips and increases elliptically, with the maximum lift occurring at the wing root.

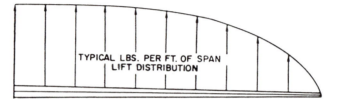

TYPICAL LBS. PER FT. OF SPAN
LIFT DISTRIBUTION

Figure 11.2   Spanwise lift distribution.

If the wing area is exactly proportional to the lift at all points on the wing, the local wing loading will be a constant value everywhere on the wing. Stall of a wing will occur where the ratio of local lift coefficient, C_l, to the wing lift coefficient, C_L, is highest. If the wing planform area is also elliptical, the local wing loading will be constant.
This is shown by the value of $C_l/C_L = 1.0$ on Figure 11.3.

Thus the entire trailing edge of an elliptical planform wing will stall at the same time. The elliptical wing is considered to be the most efficient planform and was used with much success on the British Spitfire. Structurally, the elliptical wing is difficult to manufacture and so it is not widely used.

The straight wing is usually rectangular, or it has a slight taper. The wing area near the wing tip is large compared to the lift in that area, and the wing loading is, therefore, low. Figure 11.3 shows that the value of C_l/C_L approaches zero near the wing tip for the rectangular wing. The maximum wing loading is at the wing root.

Stall will, therefore, start at the trailing edge of the wing root and then spread outward and forward as the stall progresses as shown in Figure 11.4.

Swept wings have more taper, with the extreme being the delta wing, which has a taper ratio of zero. The delta wing's area at the tip is very small, and the wing loading is very high. Stall starts where the wing loading is highest, so the stall starts at the wing tip and progresses inboard and forward. The high tip wing loading can be seen from the values of C_l/C_L in Figure 11.3. The stall pattern is shown in Figure 11.4.

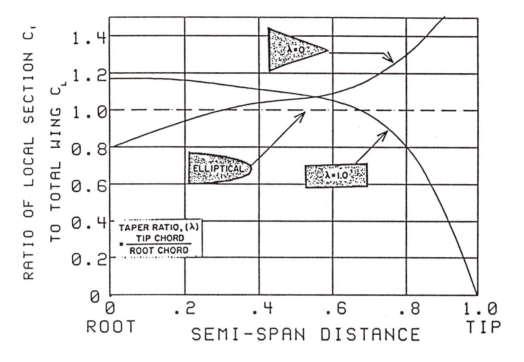

Figure 11.3   Wing spanwise lift distribution.

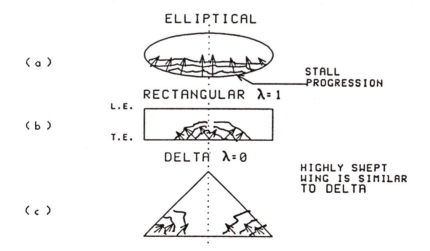

Figure 11.4   Stall Patterns.

159

Straight winged aircraft, with stall starting at the roots, seem to have advantages over swept winged aircraft, as far as stall characteristics are concerned. First, there is more adequate stall warning, caused by the separated air buffeting the fuselage. Second, the ailerons of the straight winged aircraft are not enveloped in the stalled air until the stall has progressed outward from the wing roots, compared to the relatively early envelopment from tip stalling of the swept wing. Swept winged aircraft are usually equipped with stall warning devices to compensate for the lack of buffet warning.

Despite what we have just discussed, the lack of a sharp stall points on C_L curves of these swept winged aircraft makes it nearly impossible to stall them under normal low speed flight conditions. However, this does not mean that the low speed region is safe for swept winged aircraft.

In our discussion of induced drag in Chapter Five we saw that induced drag was very high at low airspeeds. Induced drag is also greater with aircraft having low aspect ratios. Sweeping the wings reduces the span and also effectively increases the chord. Aspect ratio is the span divided by the average chord; thus swept winged aircraft, with the same wing area, have lower values of aspect ratio. The high drag existing at low airspeeds was evident in the comparison of thrust-required curves of the swept winged jet aircraft to the power-required curves of propeller driven straight winged aircraft.

Because the high drag on jet aircraft at low airspeeds is so important, a discussion of the *region of reversed command* is warranted. Most pilots are familiar with this flight region as the "backside of the thrust curve" or, incorrectly, as the "backside of the power curve." This last expression is **only correct** when propeller aircraft are being discussed.

As we have seen, propeller aircraft are not affected by a lack of power in the low speed region, so only jet aircraft and thrust required curves are considered here.

REGION OF REVERSED COMMAND

A typical thrust-required curve for a turbojet aircraft in the dirty configuration is shown in Figure 11.5. Minimum drag occurs at the $(L/D)_{MAX}$ point. For this aircraft these figures are 2400 lb. thrust required at 160 knots TAS.

All airspeeds greater than that for minimum drag are said to be in the region of normal command. In this region added thrust must be supplied if greater airspeed is desired. If the pilot of this aircraft wants to increase the airspeed to 200 knots, he must add throttle until the engine(s) produces 2750 lb. of thrust.

In the region of normal command, thrust is directly proportional to velocity.

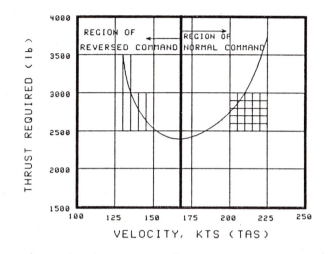

Figure 11.5   Regions of normal and reversed command.

All airspeeds below that for minimum drag are said to be in the region of reversed command. In this region, the thrust required to fly the aircraft is *inversely* related to the airspeed. The slower the aircraft flies, the greater the thrust required. At an airspeed of 140 knots, for instance, the T_R is 2750 lb., the same as was required at 200 knots.

It can be reasoned that "if more thrust is required to fly slower, then to slow down we should add throttle." This, of course, is not true. The fallacy of this statement lies in the fact that, at any stabilized airspeed, the thrust available equals the thrust required, and addition of thrust produces a thrust excess. This excess will cause the aircraft to either climb or to accelerate.

The correct statement should be, "if, for any reason, the aircraft should be slowed down, more thrust is required to maintain altitude at that airspeed." Take the case of an aircraft flying at minimum drag. This is 160 knots for the Figure 11.5 aircraft. The T_R is 2400 lb., and the T_A must also be 2400 lb. for stabilized flight. There is no excess thrust so the aircraft cannot climb under these conditions. If the pilot wants to slow the aircraft, he has several alternatives. First, he can increase the drag by using speed brakes or by dropping the landing gear. Second, he can reduce throttle setting thus reducing thrust. Third, he can increase the AOA by back pressure on the control stick.

Let us examine these alternatives in more detail. Increasing the drag is certainly a viable method of slowing the aircraft. In fact, that is why speed brakes were invented. However, speed brakes and lowering the gear usually result in rapid changes in airspeed, so let's examine how we could slowly and gradually reduce the airspeed.

Reducing the throttle results in a thrust deficiency. At any stabilized flight airspeed, a thrust deficiency results in a rate of sink. This can be seen by an adaptation of equation 6.4:

$$\text{Rate of sink} = 101.3 \ (V) \ \frac{(D - T)}{W}$$

The pilot who elects to slow the aircraft by this method cannot maintain altitude. Should he attempt to increase AOA to compensate for less thrust, he will merely decrease the airspeed, and increase the drag, resulting in an even greater thrust deficiency and greater rate of sink.

The third alternative method of slowing the aircraft was to increase the AOA. In our study of the basic lift equation, we made a point of the fact that the AOA controlled the value of the lift coefficient, and this in turn controlled the airspeed. We said, *"angle of attack is the primary control of airspeed in steady flight."*

Let's see how this works here. The pilot pulls back on the control stick and thus increases AOA. Lift coefficient is increased, airspeed is reduced, and drag is increased. A deficiency now exists in thrust, and a sink rate will develop. So, the pilot must add throttle to maintain altitude. This can be a smooth coordinated method of slowing the aircraft without loss of altitude and without excessive "throttle jockeying."

Another example of using AOA to control airspeed is illustrated in Figure 11.6.

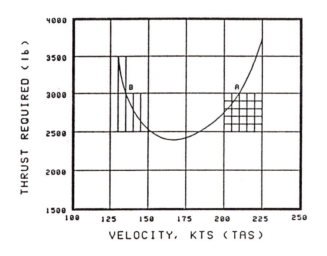

Figure 11.6   Constant airspeed climb. Stick or throttle?

162

Suppose a pilot is flying an airplane having the T_R curve shown in Figure 11.6, and the throttle is set to produce 3000 lb. of thrust. For stabilized flight, the airplane must be at point A or at point B. At point A the airspeed will be 210 knots, and at point B it will be 135 knots.

Further suppose that the plane is at point A and the pilot wants to climb the aircraft. He can pull back on the control stick and increase the AOA. This will increase the C_L and, as we have just explained, this will decrease his airspeed.

Let us say that this maneuver slows the aircraft to 200 knots. The thrust available is still 3000 lb., but the thrust required has dropped to 2750 lb. The aircraft now has an excess thrust of 250 lb. and will be able to climb.

Now, suppose the plane is at point B and the pilot wants to climb the aircraft. If he pulls back on the control stick and increases the AOA, the airspeed will be reduced again, but this time the thrust required will be increased.

For example: If the airspeed is decreased to 130 knots, the thrust required will be about 3500 lb., for a thrust deficiency of 500 lb. The climb method that worked in the region of normal command will not work in the region of reversed command.

There is a method of climbing the aircraft that works at **all** airspeeds, and it is strongly recommended. Always control the airspeed with the stick. If the airspeed is held constant, thrust required will be constant.

Climb is then controlled by the excess thrust, and the thrust available is determined by the throttle setting. So, ***throttle setting controls the rate of climb or descent in steady flight.***

In conclusion, flight in the region of reversed command cannot be avoided. Every takeoff and landing requires flight in this region. Caution must be observed so that the pilot does not "paint himself into a corner" where drag is so high that enough thrust may not be available to overcome it.

When the drag (T_R) exceeds the thrust available (T_A) there is only one way to go — **DOWN.**

LOW LEVEL WINDSHEAR

Windshear can be defined as a change in wind direction and/or speed in a very short distance in the atmosphere. While the windshear phenomenon can occur at altitude, it is most hazardous during landing and takeoff operations, so this discussion is limited to low level wind shear. Under unusual weather conditions, wind direction changes of 180° and wind speed changes as high as 50 knots have been observed.

We have been taught that wind cannot affect an aircraft inflight, except for groundspeed and drift. This is true for steady winds or winds that change gradually, but it is not true when rapid wind changes occur.

When the air velocity changes, the mass of the aircraft must be accelerated or decelerated, and this takes time.

Windshears are caused by thunderstorm activity, weather fronts, and by low level jet streams. Figure 11.7 illustrates a windshear situation caused by a thunderstorm. The downward air flow directly below the thunderstorm is called a *downburst*. Investigation of an airline crash revealed that vertical velocities in excess of 1800 fpm were encountered in such a downburst. The downburst is the most dangerous type of wind shear as it can force the airplane into the ground.

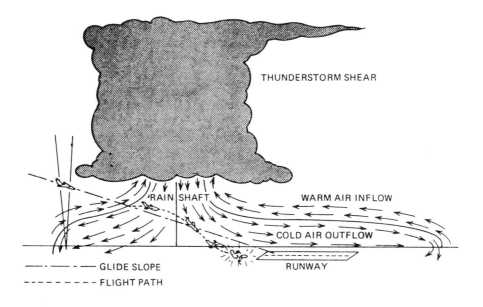

Figure 11.7 Windshear caused by a thunderstorm.

When a downburst hits the ground, it is diverted outward. This creates horizontal air velocities known as *head wind bursts*, *tail wind bursts*, and *cross wind bursts*, depending on the position of the aircraft relative to the center of the thunderstorm. These are shown in Figure 11.8, as viewed from above. The performance of an aircraft flying through these bursts is discussed in the next section.

Another hazard associated with thunderstorms is called *first gust*. This is a wind shift line or gust front that may precede a thunderstorm by as much as 15 miles. This is shown in Figure 11.9. The slope of the gust front, makes it impossible to determine when an aircraft will fly through it.

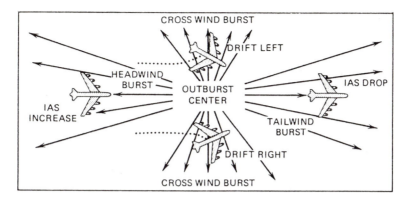

Figure 11.8   "Bursts" caused by a thunderstorm.

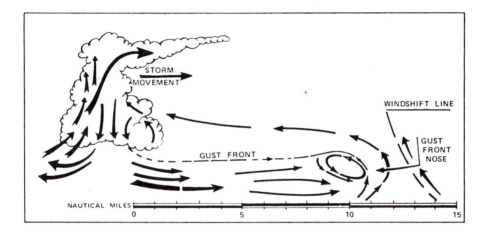

Figure 11.9   Thunderstorm gust front.

Windshear may also be present in certain types of cold and warm fronts. Turbulence may or may not exist in windshear conditions. Most fronts have shallow wind gradients; thus the changes in wind direction and velocity are gradual. Other fronts have steep wind gradients and severe amounts of windshear. A rule of thumb to determine the possibility of windshear is this: If a temperature differential of 10 °F or greater exists across the front and/or if the frontal speed is 30 knots or more, there is the possibility of significant low level windshear.

A third source of low level windshear is caused by temperature inversions. These are called *low level jet streams.*

They are caused by warm air moving above a pocket of cool, calm air, as shown in Figure 11.10. This type of windshear can be found on clear nights, when the ground is dry and smooth. Wide temperature variations during the day and night period help create this situation.

The horizontal winds at 1000 ft. above ground level have been found to be as much as 70 knots, while the surface winds remained calm.

Little or no turbulence, or vertical wind velocity, is produced inside the low level jet stream. When an aircraft is flying at a constant altitude, no significant wind shear will occur. Large variations in head wind, tail wind, and cross wind components can be experienced, however, when climbing or descending through a low level jet stream.

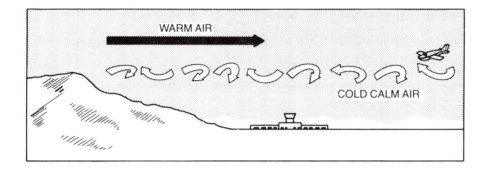

Figure 11.10   Low level jet stream.

Aircraft Performance in Low level Windshear

During Takeoff and Departure

The most hazardous condition caused by a windshear during takeoff is when an increasing tailwind is encountered shortly after takeoff as shown in Figure 11.11. The sequence of events shown is:

(1) takeoff appears normal,
(2) tailwind windshear encountered just after liftoff,
(3) airspeed decreases resulting in pitch down moment,
(4) airplane crashes.

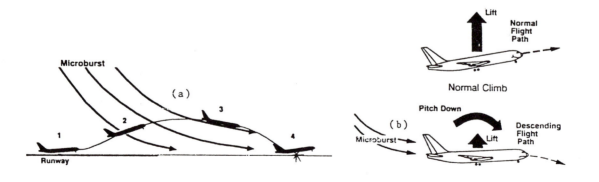

Figure 11.11   Tailwind windshear encountered on takeoff; (a) flight path, (b) forces and moments.

During Approach to a Landing

Windshear encountered during final approach can lead to dangerous situations. The two most likely conditions are discussed here.

1. If an aircraft passes through a front where a wind shift from a head wind to either a tail wind or a reduced head wind occurs, then the shearing situation shown in Figure 11.12 may exist. The sequence of events shown is:

(1) normal approach,
(2) increasing downdraft and tailwind encountered,
(3) airspeed decreases and pitch down results,
(4) aircraft crashes short of runway.

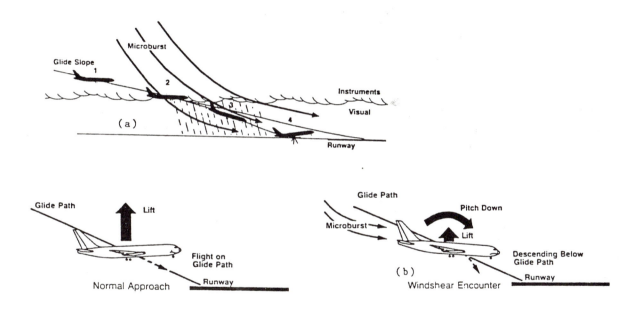

Figure 11.12   Tailwind shear encountered in landing approach; (a) flight path, (b) forces and moments.

2. When an aircraft makes an approach in a tail wind condition that changes to a head wind or calm condition, the reverse of the above situation occurs. When the windshear is encountered, the IAS increases, the aircraft pitches up, and it climbs above the glide slope. If no thrust correction is made the aircraft climbs above the flight path and a high and fast approach results with the possibility of overshooting the runway.

If the pilot reduces thrust after going high on the glide slope, he must re-establish thrust as the aircraft returns to the glide slope or the airplane will go below the flight path and a short hard landing will result.

Crosswind Burst Response

A crosswind shear burst causes the airplane to roll and/or yaw. Large crosswind shears may require large and rapid aileron inputs. These shears may increase the pilot workload and cause increased distraction. There may also be horizontal vortices present which, if encountered, may cause severe roll forces and require full aileron movement to maintain aircraft control.

Turbulence Effects

Severe turbulence may be encountered in windshear conditions. This can make changes in airspeed difficult to detect and thus delay the recognition of the presence of windshear. Turbulence may also discourage the pilot from using the pitch attitude to recover by causing stick shaker activation.

Heavy Rain

Wind shears are often accompanied by heavy rain. Earlier studies of rain effects at altitude concluded that rain had a significant effect on performance but, due to sufficient altitude, would not force the aircraft to the ground.

In the approach or post takeoff flight phase, however, conditions are much more critical and the performance margin of the aircraft is considerably reduced.

Rain affects the aircraft in several ways:

(1) The momentum of the raindrops impact the aircraft in a downward and backward direction.

(2) The increased weight of the rain water film affects the gross weight of the aircraft.

(3) The water film is roughened by the impact of the rain-drops and the aerodynamic properties of the wing are adversely affected.

(4) The raindrops hit the airplane unevenly, causing possible pitching or rolling moments.

Of the above effects, the roughened airfoil and resulting loss of aerodynamic properties is the most important. Recent studies have indicated that as much as a 30% increase in C_D and a loss of lift of more than 30% can occur due to this roughness. The AOA for $CL_{(MAX)}$ can also be reduced by 2 to 6 degrees.

At least nine major commercial aircraft accidents have been investigated and reported by the National Transportation Safety Board (NTSB) involving heavy rain during takeoffs and landings.

EFFECT OF ICE AND FROST

The formation of ice or frost on the lifting surfaces of an airplane will cause deterioration of the lifting ability of the surfaces. The effects of ice and frost on the lift coefficient curves are shown in Figure 11.13.

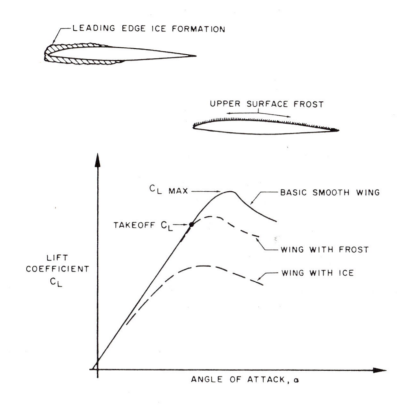

Figure 11.13   Effect of ice and frost on wings.

Ice is likely to accumulate near the leading edge if it is deposited during flight. This changes the contour of the airfoil and results in a considerable reduction of lift.

The smoothness of the skin is also reduced, and a large increase in drag results.

As a result of the reduced lift and the increased drag, thrust or power required will be greater and the stall speed will increase.

A third effect will be that the weight of the ice will increase the gross weight of the aircraft.

Obviously, all ice must be removed before flight, and any ice formed during flight must be removed by use of the aircraft's de-icing equipment. If the plane does not have de-icing equipment, a 180° turn is strongly recommended.

The effect of frost on the wings is not so obvious. A layer of evenly distributed frost will not appreciably change the contour of the wing; however, it will increase the roughness. Drag will be increased, and the kinetic energy of the air will be reduced so that stall will occur at a lower AOA and at a higher airspeed.

One study showed that 0.1 inch of evenly distributed frost on the aircraft's wings will increase the stalling speed by 35%. This roughly doubles the required takeoff run.

WAKE TURBULENCE

Turbulence caused by aircraft in flight was once attributed to "propwash." However, this phenomenon has become better understood and is now classified in two categories.

1. *Thrust stream turbulence* is one of these types. It is caused by high velocity air from propeller blades or from jet exhausts.

This type of disturbance is of primary concern during ground operations where it can be dangerous to ground personnel and to loose gear and can cause foreign object damage (FOD) to aircraft and engines.

It should not be a flight hazard except in close formation flying, or during takeoff and landing behind an aircraft that is making a ground runup.

Thrust stream turbulence dissipates fairly rapidly as compared to the second category of turbulence.

2. *Wake turbulence*, on the other hand, is generated during flight and is caused by wing tip vortices. Wing tip vortices were discussed in Chapter Five, in connection with induced drag.

Now, however, we shall discuss them as hazards to other aircraft.

A *vortex* is a highly developed rotational mass of disturbed, high energy air, created by the wing of an aircraft as it produces lift.

An aircraft creates two such rotating vortices as shown in Figure 11.14.

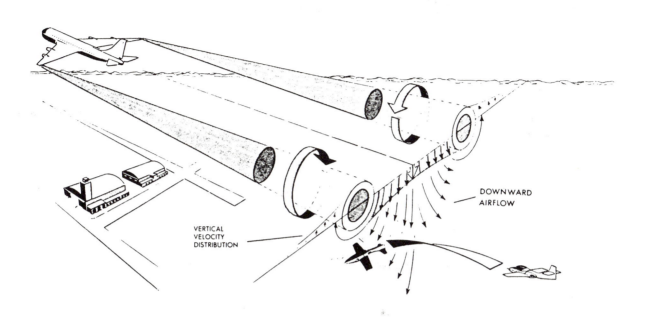

Figure 11.14   Wing tip vortices behind an aircraft.

As a lift-producing aircraft passes through the air the differential pressure, between the bottom and the top of the wing, causes a spillage of air around the wing tips. The air rolls into two distinct vortices, one at each wing tip. The rotating process is normally complete in about 2 to 4 times the wingspan (200 to 600 ft.) behind the aircraft. Vortices are also developed by helicopters and trail along behind the aircraft in the same manner as those of fixed wing aircraft. This rotational energy is directly related to the weight, lift being generated, and wingspan of the aircraft. It is inversely related to the airspeed and, therefore, the vortices are a maximum during takeoff and landing at high gross weights.

The force of the air in wing tip trailing vortices can easily exceed the aileron control capacity or climb rate of light aircraft. These forces can cause some light aircraft to roll completely over and also to be forced into the ground. A typical mishap is the case where, just before touchdown, a wing drops and contact is made with the runway while the aircraft is in a wing down attitude. Another possible result, at altitude, is the inadvertent entry into an unusual attitude and resultant spin.

Current avoidance of wake turbulence problems during takeoffs and landings is accomplished by requiring time intervals between aircraft.

Figure 11.15 shows both the takeoff and landing space clearance patterns.

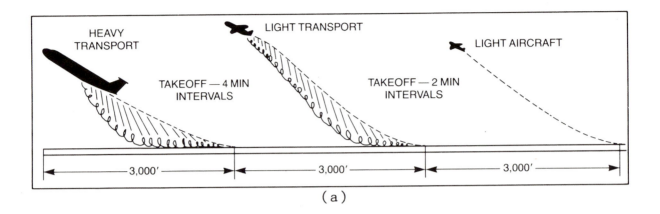

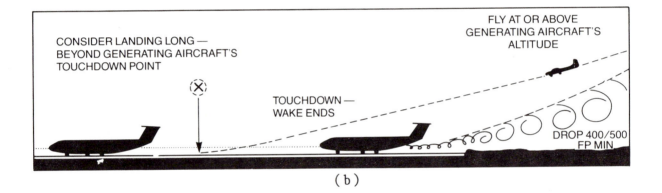

Figure 11.15   (a) Takeoff; and (b) landing clearance to avoid wake turbulence.

In the takeoff pattern the following aircraft should become airborne before the lead aircraft and should climb above the turbulence.

In the landing pattern the following aircraft should touch down past the point where the lead aircraft has touched down.

Lateral movement of tip vortices due to cross winds can present a hazard to takeoffs and landings when aircraft are operating from fields with parallel runways that are less than 2500 ft. apart. This is illustrated in Figure 11.16.

The no wind movement of the vortices is away from the centerline of the airplane at about 5 knots, as shown in Figure 11.16(a). If a 5 knot crosswind from the right is present, the left vortex will move to the left at a speed of 10 knots (Figure 11.16(b)) and will present a hazard to aircraft operating on a parallel runway to the left.

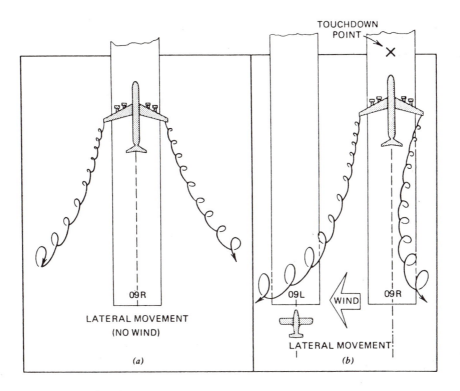

Figure 11.16 Lateral movement of tip vortices:(a) no wind; (b) with crosswind.

Much work is currently being undertaken to develop splines, winglets, and other devices to limit the amount of tip vortices. This is primarily done to reduce induced drag, but reduced wake turbulence is a spinoff advantage. In the meantime, strict adherence to existing regulations as to time and space separation is imperative.

SPINS

Intentional spins are prohibited in most light aircraft, however, spins and spin recovery should be taught (in an unrestricted airplane) as part of the training syllabus.

Spins are much more difficult to control in a jet aircraft, so they are not practiced. Actual spin recovery techniques vary for different aircraft, thus the discussion presented here is general in nature.

Spin Warning

For an aircraft to spin it must be near aerodynamic stall. Therefore stall warnings must be heeded, if spins are to be avoided. In our discussion of stalls earlier in this chapter we mentioned the advantages of the straight wing in providing stall warning to the pilot. When the pilot of a straight winged aircraft feels the buffeting, he merely has to reduce back pressure to avoid the possibility of a spin. Figure 11.17 illustrates how a straight winged aircraft stalls at the wing root first and how the stalled air envelops the tail and provides buffet warning to the pilot. At this point, the ailerons are not blanked out; thus the pilot has good lateral (roll) control.

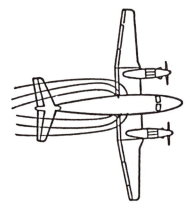

Figure 11.17   Stall affecting horizontal tail.

Swept winged aircraft stall at the wing tips first, rather than at the roots, as shown in Figure 11.18. The stall warning here is much less than for the straight winged aircraft. In addition, the ailerons are enveloped in the stalled air and become ineffective early in the stall-spin sequence. Stall warning devices are helpful in anticipating an incipient spin.

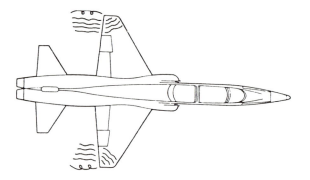

Figure 11.18   Swept wings stall at tips first.

Aerodynamic Characteristics of A Spin

The aerodynamics characteristics of a spin for a straight winged aircraft and for a swept winged aircraft are quite different.

The straight winged aircraft has C_L and C_D curves as shown in Figure 11.19.

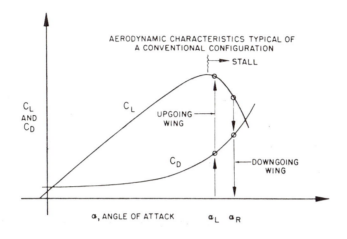

Figure 11.19   Aerodynamics of spin for straight winged aircraft.

If this type of aircraft is flying at an AOA at or above that for $C_{L(MAX)}$, and the aircraft is rolled to the right (and/or the right wing drops), the right wing is flying with a higher local AOA and becomes even more stalled.

The left wing, on the other hand, moves upward and is flying with a lower local AOA and becomes less stalled, thus producing more lift. The right rolling motion of the aircraft is aided by the relative increase in lift of the left wing, and a yawing moment to the right is produced. These moments result in autorotation.

This means that the aircraft will continue to roll and yaw without any input from the pilot. In addition to the aerodynamic forces acting on the aircraft, there are also inertia and gyroscopic effects that complicate the problem.

The straight winged aircraft's spin is characterized primarily by the rolling motion with a moderate yaw. The attitude of the spin is about 40° or more, nose down. In straight winged aircraft, a stalled condition must exist before a spin can develop, but this is not true for a swept winged aircraft.

The swept winged aircraft has C_L and C_D curves as shown in Figure 11.20. Note that the C_D curve does not have a well defined maximum lift point. When this type of aircraft is rolled at high angles of attack, only small changes in C_L take place.

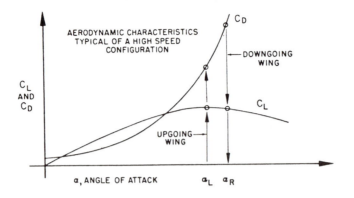

Figure 11.20 Aerodynamics of spin for swept winged aircraft.

In the swept winged aircraft there is no definite stall, and the wing autorotation contribution will be quite weak. The change in C_D that occurs between the two wings, however, is substantial, and a strong yawing moment is developed.

Swept winged aircraft have low aspect ratios. The mass of the aircraft is distributed along the longitudinal axis of the plane rather than in the wings. As the yaw develops, during the spin, this mass distribution contributes to the inertial moments and tends to flatten the spin. This results in very high angles of attack and high sink rates.

The above explanation is extremely simplified. Each part of the airplane affects pro-spin or anti-spin characteristics separately. Each model of airplane spins differently, and each will spin differently as the configuration of the aircraft is changed. The pilot must be familiar with the spin characteristics of the airplane that he flies, as described in the operator's handbook.

Spin Recovery

The most effective spin recovery technique for straight winged aircraft is to stop the spin rotation by use of opposite rudder and to lower the AOA with forward stick. Care must be used during pullout from the resulting dive to prevent an accelerated stall, which could result in an entry into another spin.

In addition to the control positions described for straight wing recovery, the ailerons are often used for swept wing spin recovery. If ailerons are used **into** the direction of the spin they create drag forces that oppose rotation. Again, consult the flight manual for proper spin recovery procedures for your aircraft.

PROBLEMS

1. The main difference between the C_L - α curves for straight winged and swept winged aircraft are:

    a. The swept wing a/c has a lower value of $C_{L(MAX)}$.
    b. The straight wing a/c does not fly at as high an AOA for $C_{L(MAX)}$.
    c. The swept wing a/c does not have as abrupt loss of lift at $C_{L(MAX)}$.
    d. All of the above.

2. Low speed stall will start at the trailing edge where:

    a. $\dfrac{C_l}{C_L}$ is a minimum.

    b. $\dfrac{C_l}{C_L}$ is a maximum.

    c. At the wing root for a swept wing.
    d. At the wing tip for a straight rectangular wing.

3. The "Region of Reversed Command" for a jet aircraft is also correctly known as:

    a. The backside of the thrust required curve.
    b. The backside of the power curve.
    c. The backside of the drag curve.
    d. Both a. and c. above.

4. An aircraft should never be flown in the "Region of Reversed Command."

    a. True.
    b. False.

5. Pulling back on the control stick (or yoke) will cause an airplane to climb at:

    a. Low speed.
    b. High speed.
    c. Any speed.

6. A better way to climb and one which will work at any speed, is to control the airspeed with the stick and add throttle to climb.

    a. True.
    b. False.

7. An airplane is making a final approach for a landing and encounters a horizontal wind shear. Which of the below types is the most dangerous?

    a. Tailwind to headwind shear.
    b. Downburst.
    c. Headwind to tailwind shear.
    d. Crosswind burst.

8. An airplane in flight encounters icing. The greatest danger is that:

    a. Weight is increased.
    b. Drag is increased.
    c. Lift is decreased.

9. Wake turbulence can cause an airplane to be turned completely upside down. To avoid wake turbulence a pilot should avoid:

    a. Flying behind and below a large airplane.
    b. Taking off or landing behind a large airplane.
    c. Taking off or landing on a parallel runway, downwind from that being used by large airplanes.
    d. All of the above.

10. Spin recovery consists of:

    a. Stopping spin rotation with rudder.
    b. Lowering the AOA with forward stick.
    c. Applying ailerons into the spin (swept wing a/c).
    d. All of the above.

11. A 10,000 lb. airplane has the thrust required curve shown in Figure 11.6. The thrust available is 3000 lb. and is considered to be the same at all airspeeds. The airplane is flying at 210 knots and the pilot decides to climb and pulls back on the stick, thus reducing the airspeed to 200 knots. What will the rate of climb (or descent) be?

12. The same situation exists as in problem 11, except the airspeed at the start of the maneuver is 135 knots and 130 knots after the pilot pulls back on the stick. What will the rate of climb (or descent) be?

CHAPTER TWELVE

TAKEOFF PERFORMANCE

Takeoff performance involves accelerated motion. During the takeoff, the aircraft starts from zero velocity, accelerates to takeoff velocity, and becomes airborne.

The relationships among acceleration, velocity, takeoff distance, and time are not familiar to the average pilot and should be understood as well as experienced. This chapter discusses the factors involved in attaining takeoff velocity.

The following are important factors in takeoff performance:

1. Takeoff velocity. This could be a function of stall speed, minimum control speed, or the thrust (or power) available-thrust (or power) required relationship.

2. The acceleration during takeoff.

3. The distance required to complete the takeoff.

Takeoff and landing distance computation is a complex subject, and the takeoff charts in the pilot's handbook should be closely studied. This book makes no effort to analyze the problems in detail but does attempt to explain, in a simplified manner, how changes in weight, altitude, runway slope, head wind/tail wind, and other factors influence the aircraft's performance.

The problem is simplified if we assume that the aircraft's acceleration is a constant during the takeoff roll.

LINEAR MOTION

Newton's laws of motion express relationships among force, mass, and acceleration, but they stop short of discussing velocity, time and distance. These are covered here. The equations that are derived here are only valid for constant acceleration.

$$\text{Acceleration, } a = \frac{\text{change in velocity}}{\text{change in time}} = \frac{V - V_o}{t - t_o}$$

where
   V = velocity at time t
   V_o = velocity at time t_o

If we start the time at $t_o = 0$ and rearrange the above:

$$V = V_o + at$$

(12.1)

The distance, s, traveled in a certain time is

$$s = (V_{AV})\,(t)$$

The average velocity, V_{AV}, is:

$$V_{AV} = \frac{1}{2}(V + V_o)$$

So:

$$s = \frac{1}{2}(V_o + at + V_o)t \;\; \text{or,} \;\; s = V_o t + \frac{1}{2}at^2$$

(12.2)

Solving equations 12.1 and 12.2 simultaneously and eliminating t, we can derive a third equation:

$$s = \frac{(V^2 - V_o^2)}{2a}$$

(12.3)

In applying these equations to the takeoff problem where the aircraft starts from a "brakes locked" position on the runway, V_o is zero. The equations are then simplified and equation 11.3 becomes:

$$s = \frac{V^2}{2a}$$

(12.4)

where
   s = distance (ft)
   V = takeoff velocity (fps)
   a = acceleration (fps^2)

The acceleration of an aircraft during takeoff is determined by applying Newton's second law: F = ma. The force that provides the acceleration for takeoff is the unbalanced force acting on the aircraft during the takeoff roll. This is called the net accelerating force and is illustrated in Figure 12.1. There are two assumptions made in constructing Figure 12.1. The first of these is that there is no lift being developed as the aircraft gathers speed. This is true for many of the higher performance aircraft which have the wing positioned on the aircraft at an angle of incidence which produces the least drag rather than one which produces lift. These aircraft must "rotate" at takeoff speed to develop lift. Most transport aircraft and military aircraft fit into this category. If no lift is developed during the takeoff run, rolling friction remains constant until rotation. The second assumption is that the thrust increases during acceleration (which it does) and that net accelerating force is constant throughout the takeoff run and thus equation 12.4 is valid.

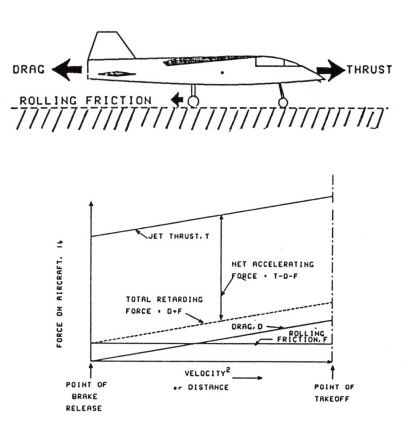

Figure 12.1 Forces on an airplane during takeoff.

Figure 12.1 shows that the net accelerating force on the aircraft, F_N, equals the thrust T, minus the drag D, minus the rolling friction, F:

$$F_N = T - D - F$$

By substituting in Newton's second law equation and rewriting we obtain:

$$a = \frac{F_N}{m} = \frac{g(T - D - F)}{W}$$

(12.5)

where:
   a = acceleration (fps^2)
   F_N = net accelerating force (lb)
   W = weight (lb)
   g = gravitational acceleration (= 32 fps^2)
   m = mass, slugs (= $\frac{W}{g}$)

Equation 12.5 shows that the acceleration of the aircraft is affected by the thrust, the drag, the rolling friction, and the weight of the aircraft.

FACTORS AFFECTING TAKEOFF PERFORMANCE

Factors that affect takeoff performance are:

1. Aircraft gross weight.
2. Thrust on the aircraft.
3. Temperature.
4. Pressure altitude.
5. Wind direction and velocity.
6. Runway slope.
7. Runway surface.

To see how these factors affect takeoff distance, equation 12.4 is used to make a ratio. Subscripts 1 denote a known set of conditions, and subscripts 2 apply to a new set of conditions.

$$\frac{s_2}{s_1} = \left(\frac{V_2}{V_1}\right)^2 \left(\frac{a_1}{a_2}\right)$$

$$(12.6)$$

The effects of weight change, altitude change, and wind conditions on takeoff distance can be found by applying equation 12.6.

Effect of Weight Change

Increasing the gross weight of an aircraft has a threefold effect on takeoff performance: (1) takeoff velocity is increased, (2) there is more mass to accelerate and, (3) there is more rolling friction.

Takeoff velocity varies as the square of the weight. This was discussed in Chapter Four and shown in equation 4.2:

$$\frac{V_2}{V_1} = \sqrt{\frac{W_2}{W_1}}$$

Squaring this equation:

$$\left(\frac{V_2}{V_1}\right)^2 = \frac{W_2}{W_1}$$

The effect of increasing weight on an aircraft's acceleration is twofold. There is more mass to be accelerated, and there is more rolling friction. For a dry concrete runway, the coefficient of rolling friction is about 0.03, and therefore an increase in weight of 1000 lb means a loss in accelerating force of 30 lb. This change is small, so the principal effect on acceleration is due to the change in the mass of the aircraft.

Acceleration is inversely proportional to the mass (or weight) of the aircraft. Therefore, if the rolling friction increase is ignored, the acceleration term is:

$$\frac{a_1}{a_2} = \frac{W_2}{W_1}$$

If we substitute values from these last two equations into equation 12.6, the effect of a weight change on takeoff distance is:

$$\frac{s_2}{s_1} = \left(\frac{W_2}{W_1}\right)^2 \tag{12.7}$$

Effect of Altitude

Pressure altitude corrected for runway temperature results in runway density altitude. This may be considerably higher than the field density altitude, because runway temperatures can be higher than the official field temperature.

An increase in density altitude has a twofold effect on takeoff performance.

First, a higher takeoff velocity (TAS) is required.
Second, for unsupercharged reciprocating engines and for turbine engines, less thrust is available.

To produce lift equal to the weight of the aircraft, at a given AOA, the dynamic pressure must be the same, regardless of the altitude. Thus the airplane will take off at the same equivalent airspeed (EAS) at altitude as at sea level. Because of the reduced density of the air, the true airspeed (TAS) will, however, be greater at altitude. From basic aerodynamics, we learned that the relationship between true and equivalent airspeed was expressed by equation 2.12:

$$\frac{TAS}{EAS} = \frac{1}{\sqrt{\sigma}}$$

Thus the velocity for takeoff at an airfield with a density altitude above sea level, (V_2), as compared to the velocity required at sea level on a standard day, (V_1) is:

$$\frac{V_2}{V_1} = \frac{1}{\sqrt{\sigma_2}} \text{ or } \left(\frac{V_2}{V_1}\right)^2 = \frac{1}{\sigma_2}$$

For unsupercharged reciprocating engines and turbine engines, the power or thrust available decreases approximately as the air density (or density ratio). As thrust is decreased, acceleration, (a_2), is also decreased, as follows:

$$\frac{a_1}{a_2} = \frac{1}{\sigma_2}$$

Substituting values from these last two equations into equation 12.6 gives us the effect of altitude on takeoff distance for unsupercharged reciprocating or turbine engine aircraft:

$$\frac{s_2}{s_1} = \left(\frac{1}{\sigma_2}\right)^2 \qquad (12.8)$$

Supercharged reciprocating engines can deliver sea level power up to their critical altitude, so no decrease in thrust or acceleration occurs below that altitude, thus:

$$\frac{s_2}{s_1} = \frac{1}{\sigma_2} \qquad (12.9)$$

where:

s_1 = standard sea level takeoff distance (ft.)

s_2 = altitude takeoff distance

σ_2 = altitude density ratio

Effect of Wind

Takeoff into a head wind allows the aircraft to reach takeoff velocity at a lower groundspeed than for a no-wind condition. The takeoff velocity, V_2, with a head wind, in terms of the no-wind takeoff velocity, V_1, and the velocity of the head wind V_w, is:

$$V_2 = V_1 - V_w$$

Dividing both sides of this equation by V_1 and squaring:

$$\left(\frac{V_2}{V_1}\right)^2 = \left(\frac{V_1 - V_w}{V_1}\right)^2 = \left(1 - \frac{V_w}{V_1}\right)^2$$

Although the acceleration over the ground is reduced by a head wind, the acceleration through the air mass is not. Therefore:

$$a_2 = a_1 \quad \text{or} \quad \frac{a_1}{a_2} = 1$$

Substituting the values of these last two equations into equation 12.6, we find that the effect of a head wind on the takeoff distance is:

$$\frac{s_2}{s_1} = \left(1 - \frac{V_w}{V_1}\right)^2 \qquad (12.10)$$

Takeoff with a tail wind simply means that the groundspeed must be increased over the no-wind groundspeed by the amount of the tail wind.

$$\frac{s_2}{s_1} = \left(1 + \frac{V_W}{V_1}\right)^2$$

(12.11)

Effect of Runway Slope

If the runway has a slope to it, the component of the weight that is parallel to the runway surface will reduce the net accelerating force, in case of an upslope, or increase the accelerating force, in case of a downslope:

Accelerating force = ± W sine slope°

Pilot's handbook takeoff distance charts usually include runway slope calculations. A rule of thumb is "Five percent increase in takeoff distance for each percent of uphill slope."

ABORTED TAKEOFFS

Definitions

Runway Available. Actual runway length less overrun and the airplane line-up distance (200 ft.).

Takeoff Speed, (V_2). The speed to which the airplane must be accelerated before leaving the runway.

Takeoff Ground Run. The distance through which the airplane must be accelerated, without loss of an engine, to reach takeoff speed.

Critical Engine Failure Speed, (V_{CEF}). The speed to which the airplane can be accelerated, lose an engine, and then either continue to the takeoff with the remaining engines or stop, in the same total runway distance. V_{CEF} does not apply to single engine aircraft.

Critical Field Length, (CFL). The total length of runway required to accelerate on all engines to critical engine failure speed, experience an engine failure, and then continue to takeoff or stop. For a safe takeoff the CFL must be no greater than the runway available. CFL does not apply to single engine aircraft.

Ground Minimum Control Speed, (V_{MGC}). The minimum airspeed at which the airplane, while on the ground, can lose an outboard engine and maintain directional control. V_{MGC} does not apply to single engine aircraft.

In-Flight Minimum Control Speed, (V_{MCA}). The minimum speed at which an engine can be lost and directional control maintained using full rudder deflection and not more than 5° of bank. V_{MCA} does not apply to single engine aircraft.

Refusal Speed, (V_{REF}). The maximum speed that the aircraft can obtain under normal acceleration and then stop in the available runway.
V_{REF} does apply to single engine as well as to multiengine aircraft.

Maximum Braking Speed, (V_{MB}). The highest speed from which the airplane may be brought to a stop without exceeding the maximum energy absorption capacity of the brakes.

Takeoff Decision Speed, (V_1). The minimum indicated airspeed at which an engine failure can be experienced and the takeoff safely continued.
   Safe abort capability is assured if the takeoff is aborted prior to reaching this speed.
   V_1 is the critical engine failure speed or ground minimum control speed, whichever is higher. However, it must never exceed refusal speed, maximum braking speed, or takeoff speed.
   V_1 does not apply to single engine aircraft.
   A graphic representation of the refused takeoff situation is shown in Figure 12.2.

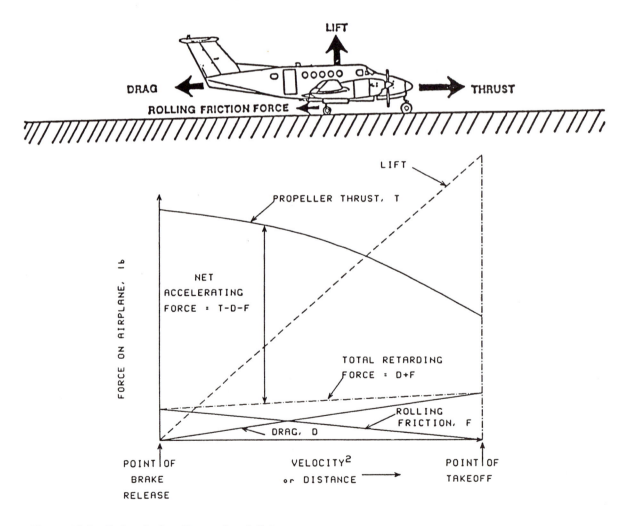

Figure 12.2   Refusal takeoff speed and distance.

186

At the brakes release point the aircraft starts to accelerate, and the variation of velocity and distance is as shown by the takeoff acceleration profile. The deceleration profile shows the variation of velocity with distance as the airplane is brought to a stop at the end of the runway.

The refusal speed is determined by the intersection of the acceleration and deceleration profiles. An allowance for pilot reaction time is included in the calculation of refusal speeds. Once the refusal speed is exceeded, the aircraft cannot be brought to a stop in the runway remaining.

If a single engine airplane loses its engine above refusal speed it will end up in the overrun.

A multiengine airplane should continue to make an engine-out takeoff if an engine fails above this airspeed.

This is illustrated in Figure 12.3.

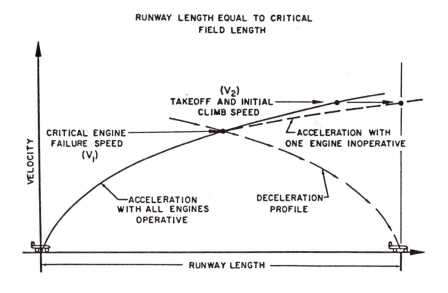

Figure 12.3   Multiengine velocity-distance profiles.

The refusal speed for multiengine aircraft is called the *critical engine failure speed*, (V_{CEF}). If the airplane has not reached this speed, the pilot can retard all engines and brake the aircraft to a stop before the end of the runway in the same manner as for the single engine airplane.

At or above V_1 the pilot of the multiengine airplane must continue the takeoff after losing an engine. As can be seen in Figure 12.3, acceleration will be less with the loss of an engine, but takeoff velocity, V_2, will be the same. Thus takeoff distance will be greater. In some cases, the remaining runway may not be enough to allow the airplane to accelerate to V_2. Thus critical field length becomes important since it assures that this will not happen.

Figure 12.3 shows the condition where actual field length is exactly equal to critical field length and V_2 is reached at the exact end of the runway.

SYMBOLS AND UNITS

English Symbols

F	Rolling friction force (lb.)
F_N	Net accelerating force
V_1	Takeoff decision speed (knots)
V_2	Takeoff speed
V_{CEF}	Critical engine failure speed (V_1 or V_{MCG})
V_{MB}	Maximum braking speed
V_{MCA}	In-flight minimum control speed
V_{MGC}	Ground minimum control speed
V_{REF}	Refusal speed
V_W	Wind speed

EQUATIONS

12.1 $\quad V = V_o + at$

12.2 $\quad s = V_o t + \dfrac{1}{2}at^2$

12.3 $\quad s = \dfrac{(V^2 - V_o^2)}{2a}$

12.4 $\quad s = \dfrac{V^2}{2a}$

12.5 $\quad a = \dfrac{g(T - D - F)}{W}$

12.6 $\quad \dfrac{s_2}{s_1} = \left(\dfrac{V_2}{V_1}\right)^2 \left(\dfrac{a_1}{a_2}\right)$

12.7 $\quad \dfrac{s_2}{s_1} = \left(\dfrac{W_2}{W_1}\right)^2$

12.8 $\quad \dfrac{s_2}{s_1} = \left(\dfrac{1}{\sigma_2}\right)^2$ (unsupercharged reciprocating or turbine engined aircraft)

12.9 $\quad \dfrac{s_2}{s_1} = \dfrac{1}{\sigma_2}$ (supercharged reciprocating engined aircraft below critical altitude)

12.10 $\quad \dfrac{s_2}{s_1} = \left(1 - \dfrac{V_W}{V_1}\right)^2$ (headwind)

12.11 $\quad \dfrac{s_2}{s_1} = \left(1 + \dfrac{V_W}{V_1}\right)^2$ (tailwind)

PROBLEMS

1. Takeoff velocity for a multiengine airplane is a function of:

    a. Stall speed.
    b. Minimum control speed.
    c. Low speed region where $T_A = T_R$ or $P_A = P_R$.
    d. The highest of the above.

2. Takeoff distance is a function of:

    a. Takeoff velocity.
    b. Acceleration.
    c. Both of the above.

3. Takeoff distance is:

    a. Directly proportional to the acceleration.
    b. Inversely proportional to the velocity squared.
    c. Inversely proportional to the acceleration.
    d. Directly proportional to the velocity squared.
    e. Both c. and d. above.

4. As weight increases the takeoff distance increases because:

    a. there is less thrust and thus less acceleration.
    b. there is more mass and thus less acceleration.
    c. the takeoff velocity is higher.
    d. both b. and c. above.

5. As weight (W_2) increases the takeoff distance increases:

    a. Directly as the square of W_2/W_1.
    b. Directly as W_2/W_1.
    c. Inversely as the square of W_2/W_1.
    d. Inversely as W_2/W_1.

6. For jet aircraft, takeoff distance at altitude is greater than at sea level because:

    a. The thrust available is less.
    b. The EAS for takeoff is greater.
    c. The TAS for takeoff is higher.
    d. Both a. and c. above.

Takeoff Performance

7. Which of the below aircraft will pay the greatest penalty in high altitude takeoff distance?

    a. One with a supercharged reciprocating engine.
    b. One with a turbine engine.
    c. One with an unsupercharged reciprocating engine.
    d. Both b. and c. above.

8. Takeoff acceleration through the air mass is not reduced by a headwind.

    a. True.
    b. False.

9. Takeoff with a tail wind requires that the takeoff groundspeed be _____ by the amount of the tail wind.

    a. Increased.
    b. Decreased.
    c. Not affected.

Aircraft data for problems 10-15:

    Thrust available during takeoff = 3000 lb.
    Turbo-jet engines
    Average drag + rolling friction = 500 lb.
    Gross weight = 10,000 lb.
    Takeoff speed 120 knots (203 fps)
    Sea level standard conditions.

10. Calculate the airplane's acceleration.

11. What should his airspeed be at the 1000 ft. runway marker?

12. Calculate the no-wind takeoff distance.

13. If the weight is increased to 15,000 lb., calculate the no-wind takeoff distance.

14. Calculate the takeoff distance for the 10,000 lb. airplane if there is a 12 knot headwind.

15. Calculate the no-wind takeoff distance for the 10,000 lb. airplane operating from an airfield where the density ratio is 0.8.

190

CHAPTER THIRTEEN

LANDING PERFORMANCE

Landing performance involves decelerated motion. During the landing phase of flight, the aircraft touches down at a certain velocity, then decelerates to zero velocity. This chapter discusses the approach to a landing and factors in reducing speed once the aircraft has touched down. The following are important factors in landing performance:

1. Approach paths and approach speeds.

2. Hazards of hydroplaning.

3. Deceleration during landing.

4. The distance required to stop the aircraft.

As we discussed in the previous chapter, this is a complex subject and the simplified discussion presented here is in no way to be used to supersede the information presented in the pilot's handbook.

PRE-LANDING PERFORMANCE

Gliding Flight

One of the most important concepts of aerodynamics is "equilibrium of forces." This was discussed in Chapter One. If you do not thoroughly understand this concept, you should review that material now. An airplane that is flying at a constant speed is in equilibrium, as shown in Figure 1.2. If the engine(s) should fail, the thrust force is reduced to zero, and an unbalance of forces results. This means that the aircraft will not be able to maintain altitude and will start to descend. Should this occur the pilot would want to know several things:

1. How far can the airplane glide?

2. How long will the airplane remain airborne?

3. What will the sink rate be?

4. Can the plane glide to a suitable landing site?

5. Can a successful engine-out landing be made?

An understanding of the basic aerodynamics of gliding flight should help the pilot make an intelligent estimate of the performance of the airplane in the above situation. Once the engine quits and the aircraft starts to descend, the forces acting on it are as shown in Figure 13.1.

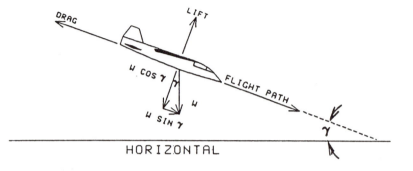

Figure 13.1 Forces acting in a power-off glide.

The figure shows that the forces of lift and drag act along the vertical and longitudinal axes of the aircraft, but the weight acts toward the center of the earth. To better analyze the reaction of these forces, it is desirable to replace the weight force by two component forces that act in the same rectangular coordinate system as the lift and drag forces. The glide path makes a glide angle with the horizontal which we will call γ (gamma).

The component of the weight that acts in the direction of the aircraft's vertical axis is equal to W cos γ. It acts at the aircraft's center of gravity (CG) and opposes the lift L.

The component of weight that acts along the longitudinal axis is equal to W sin γ and opposes the drag D. In Figure 13.1 the component W sin γ is shown at some distance below the CG. This component actually acts at the CG and is merely shown in the figure to explain the right triangle relationship of the vectors.

Once glide has been established and velocity is constant, equilibrium again exists and the force equations are

$$L = W \cos \gamma \tag{13.1}$$

$$D = W \sin \gamma \tag{13.2}$$

The pilot is concerned about how to achieve maximum glide ratio. Flying at maximum glide ratio means that maximum glide range will be attained.

The ratio of horizontal distance to the vertical distance (altitude) is the *glide ratio*. To cover the most distance over the ground, the pilot wants to have the maximum glide ratio. The flight path angle must be a minimum to achieve this.

Dividing equation 13.1 by 13.2 gives:

$$\frac{L}{D} = \frac{W \cos \gamma}{W \sin \gamma} = \frac{1}{\tan \gamma} \quad \text{or} \quad \tan \gamma = \frac{D}{L} = \frac{1}{L/D} \tag{13.3}$$

Equation 13.3 tells us that the minimum glide angle is obtained at $(L/D)_{MAX}$ and, therefore, the airplane must be flown at the airspeed corresponding to the $(L/D)_{MAX}$ AOA. The glide ratio vector diagram is shown in Figure 13.2.

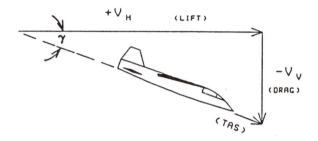

Figure 13.2   Glide ratio vector diagram.

The tangent of the glide angle in Figure 13.2 is equal to the opposite side divided by the adjacent side of the right angle shown in the figure.

From equation 13.3, the tangent of the glide angle was D/L, so the opposite side of the triangle can be labeled as (DRAG) and the adjacent side can be labeled as (LIFT). The vertical velocity $-V_V$ and the horizontal velocity $+V_H$ also represent the opposite side and adjacent side of the triangle respectively, thus the numerical value of the L/D ratio is equal to the glide ratio.

A pilot who tries to stretch the glide by flattening the glide angle will actually decrease the glide distance. Maximum glide distance is achieved only at a minimum glide angle, and this occurs when the plane flies at $(L/D)_{MAX}$.

The Landing Approach

Precision flying is required to ensure a stabilized steady flight path to touchdown. The approach speed specified for the weight of the aircraft is given in the pilot's handbook and provides a margin of safety above the minimum safe airspeed for the aircraft. The minimum airspeed may depend on the aircraft's stall speed, minimum control speed, or the speed at which $P_A = P_R$ (or $T_A = T_R$).

A fairly long approach path allows the pilot to stabilize the variable quantities such as airspeed, glide slope, and drift. Once these are brought under control, a smooth flight path can be maintained, and a minimum of control forces will be needed to make the actual landing.

Steep turns should be avoided during the alignment of the aircraft with the runway. Steep turns increase the induced drag and will bleed off the airspeed. Stall speed also increases in the turn.

Landing Performance

Once proper alignment to the runway has been established, only minor banking for drift corrections will be required. The major task now is flying the proper approach glide path. Various approach glide paths are depicted in Figure 13.3.

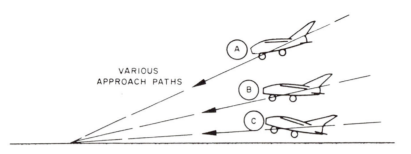

VARIOUS
APPROACH PATHS

Figure 13.3   Approach glide paths.

Approach path Ⓐ shows a steep, low power approach. Such an approach usually results from making the turn from the downwind leg too soon and not allowing for a long enough straight-away. The pilot finds the plane too close to the touchdown point and at too high an altitude. There are three alternatives:

1. Cut the throttle and attempt to lose altitude rapidly. If the approach is made at idle throttle, dive (or sideslip) the airplane to lose altitude.

2. Continue the normal approach and thus land long on the runway.

3. Execute a go-around and make a new approach.

The first alternative results in a high rate of descent and/or an increase in airspeed. Pilots of light aircraft may not have difficulty in reducing the high rate of descent by either flaring the airplane or adding throttle.

In either case the plane will probably make a hard landing or float for a distance. The high rate of descent is much more serious for heavier airplanes. The recovery involves more than merely pulling the nose up to flare the plane.

Such a recovery technique will add greatly to the induced drag, but will provide little increase in lift. A large increase in thrust is required to overcome the high rate of descent. In some cases there is simply not enough thrust available and a hard, short landing results.

The second alternative is to land long, and it may succeed if the runway is long enough.

The third alternative is to make a go-around. This is recommended and, although it may be hard on pilot ego, it is easy on the airplane. If the approach is not properly lined up on final approach, it is certainly better to "take it around" rather than to "press on" and hope to salvage a good landing from a poor approach.

Approach path Ⓒ in Figure 13.3 shows a long, shallow approach. This is a result of getting too low during the approach and then having to "drag it in" using high thrust and high AOA. Such approaches have been successful in propeller aircraft, but they must be avoided in turbojets. Propeller aircraft derive much more lift at high power settings due to the air being blown over the wing area behind the propellers, as shown in Figure 13.4(a).

The only advantage of making a high thrust approach in a conventional jet aircraft is the generation of lift from the vertical component of thrust, as shown in Figure 13.4(b). The large increase in drag that results from the high AOA for the jet aircraft far exceeds the benefits of the lift.

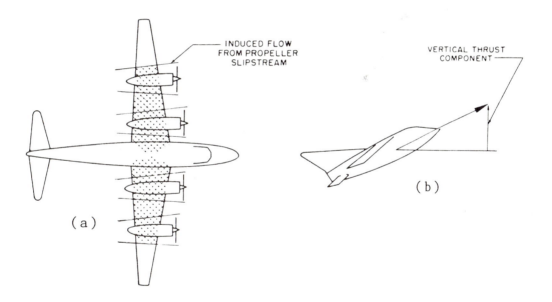

Figure 13.4   Lift from: (a) propellers; (b) turbojets.

Propeller aircraft are power producers and, as shown from equation 1.6, this means that they produce high thrust at low airspeeds. Propeller aircraft do not suffer from a thrust deficiency at low airspeeds as do jet aircraft. Even with the advantages of blowing air and high thrust available, propeller aircraft should not be flown in a low, flat, high powered approach.

Engine failure under this type of approach can be disastrous.

Thus the intermediate approach Ⓑ is desirable. It requires neither a high rate of sink, excessive speed, nor the extreme flare that the steep approach does. The intermediate path also does not require the high thrust and AOA that the low path does.

Once established on the proper approach path and aircraft heading, the pilot must control his airspeed and altitude. The proper AOA will produce the desired airspeed. Lowering the AOA will increase the airspeed, and increasing the AOA will lower the airspeed. As we have emphasized several times, *the stick controls the airspeed.* Once the proper airspeed has been established, *the primary control of rate of descent will be the throttle setting.*

LANDING DECELERATION, VELOCITY, AND DISTANCE

Forces on the Aircraft During Landing

Transport, and other large aircraft, use a different landing technique than fighter and attack type aircraft. Nearly all modern transports have non-skid brakes and thrust reversers. They land in a three point attitude and apply brakes and thrust reversers soon after touchdown. The landing forces on these aircraft are shown in Figure 13.5. Thrust reverser forces are not shown in the figure.

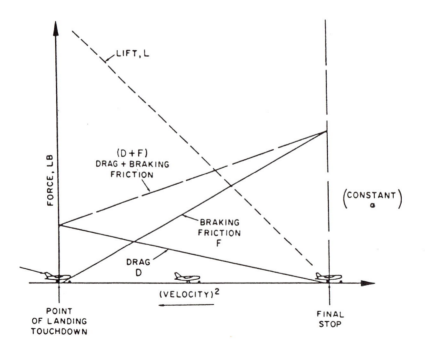

Figure 13.5   *Forces acting on an airplane during landing.*

Fighter, attack, and similar high performance aircraft use aerodynamic drag braking as well as brake friction to slow the airplane.

These forces are shown in Figure 13.6. It is assumed that:

1. The aircraft has a tricycle landing gear and lands on the main gear and the pilot hold the nose up to take advantage of the aerodynamic drag.
2. The pilot does not apply wheel brakes until he can no longer keep the nose wheel off the runway.
3. The aircraft does not develop lift in the three point position and the nose wheel is on the runway.

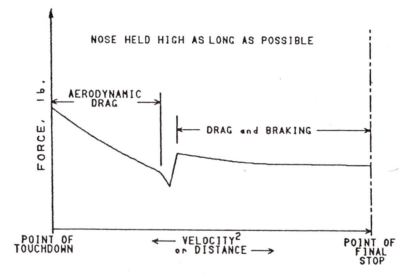

Figure 13.6   Aerodynamic braking and wheel braking.

High performance aircraft often employ drag chutes to increase the aerodynamic drag in landing. This force is not illustrated in Figure 13.6.

Braking action of the wheel brakes is not well understood by the average pilot. Many factors are involved, including:

(a) tire material,
(b) tread design and condition,
(c) runway surface material and condition,
(d) amount of braking applied (wheel slippage) and,
(e) amount of normal (squeezing) force between tires and the runway.

Friction is defined as a force that develops between two surfaces that are in contact with each other when an attempt is made to move them relative to each other. The friction force F is related to the force that is pressing the surfaces together, called the normal force, N, by a dimensionless factor called the *coefficient of friction*, μ (mu).

This is expressed by:

$$F = \mu N$$

Figure 13.7 illustrates these forces.

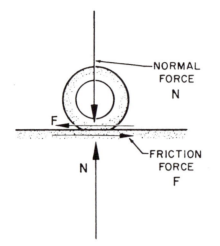

Figure 13.7   Normal and friction forces.

The coefficient of friction for a certain tire material, tread design, and tire wear operating on a dry concrete runway is a function of the amount of brakes being applied. This braking is defined as a percentage of slip between the tire and the runway. Zero percent slip is a rolling tire with no braking action. A locked wheel that skids without turning is defined as 100 percent slip. A plot of the coefficient of friction versus wheel slippage is shown in Figure 13.8.

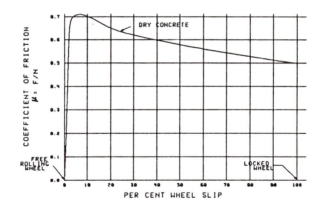

Figure 13.8   Coefficient of friction versus wheel slippage.

The maximum braking coefficient of friction can be seen to be about 0.7. It occurs when wheel slip is about 10 percent. Locking the wheels reduces the value to about 0.5, causes the tires to wear unevenly, and may possibly cause tire blowout.

Water, snow, or ice on the runway causes a large reduction of the coefficient of friction, as is shown in Figure 13.9.

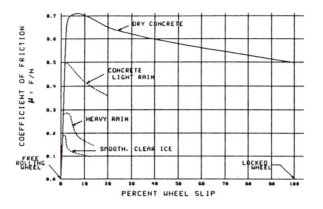

Figure 13.9   Effect of runway condition on coefficient of friction.

Braking Techniques

The technique of using braking devices may vary in detail with aircraft type and model, but general principles can be stated for aircraft using aerodynamic braking.

During the initial landing roll the velocity and dynamic pressure is high and aerodynamic braking will be most effective. Holding the nose of a tricycle landing gear airplane high will increase the parasite drag and is an effective method of slowing the aircraft. Use of speed brakes and full down flaps has the same effect.

At some point during the landing roll the reduced airspeed will cause the elevator to lose effectiveness, and the nose of the aircraft can no longer be held off the runway. Usually this is the point where wheel braking is initiated and becomes the principal method of stopping the aircraft.

There is often a debate over whether it is better to retract the flaps, thus increasing the weight on the wheels, or to leave them down, thus preserving the aerodynamic braking effect.

Inadvertent retraction of the landing gear instead of the flaps has occurred and may influence this decision.

Wheel braking force for all aircraft will be a maximum when the normal force on the braked wheels is greatest. For a tricycle landing gear aircraft, this will occur when the control stick is held in the back position.

Many pilots have a tendency to release back pressure when the nose wheel contacts the runway. This should be avoided, and the stick should be held full back as long as braking action is desired.

LANDING EQUATIONS

General Equation

The calculation of landing distance, s, in terms of touch down velocity, Vo, and deceleration, (–a), is similar to the takeoff case, which was shown in equation 12.4 to be:

$$s = -\frac{V_o{}^2}{2(-a)} = \frac{V_o{}^2}{2a}.$$

To determine how the landing distances are affected by variations in weight, altitude, and wind conditions, a ratio is established similar to that used for the takeoff case. This equation was previously numbered as equation 12.6:

$$\frac{s_2}{s_1} = \left(\frac{V_2}{V_1}\right)^2 \left(\frac{a_1}{a_2}\right)$$

Effect of Weight Change

Changing the gross weight of a landing aircraft affects the landing speed in the same way the takeoff speed was affected. The landing speed must be changed by using equation 4.2:

$$\frac{V_2}{V_1} = \sqrt{\frac{W_2}{W_1}} \quad \text{or} \quad \left(\frac{V_2}{V_1}\right)^2 = \frac{W_2}{W_1}$$

The effect of changing the weight of the aircraft has no effect on the deceleration of the aircraft, caused by the braking action. At first, this statement sounds strange, but the heavier the aircraft, the more the weight on the braking wheels. So, more braking force is available to counteract the greater weight to be decelerated. For the braked phase of the landing:

$$\frac{a_1}{a_2} = 1$$

Substituting these last two equations into equation 12.6, the effect of a weight change on landing distance is:

$$\frac{s_2}{s_1} = \frac{W_2}{W_1}$$

(13.4)

Effect of Altitude

Aircraft airspeeds are affected by the same factors at altitude whether the aircraft is taking off or landing. Thus the equation showing the velocity change at altitude holds for the landing phase as well:

$$\left(\frac{V_2}{V_1}\right)^2 = \frac{1}{\sigma_2}$$

Unless the aircraft is equipped with thrust reversers, it does not rely on the engine performance to decelerate the aircraft as it did for takeoff acceleration. So:

$$\frac{a_1}{a_2} = 1$$

Substituting the values from these last two equations into equation 12.6 gives us the effect of altitude on landing distance for all non-thrust reverser aircraft

$$\frac{s_2}{s_1} = \frac{1}{\sigma_2} \tag{13.5}$$

Effect of Wind

Head winds and tail winds affect takeoff and landing performance in exactly the same way, so equations 12.10 and 12.11 apply to the landing situation also. We will rewrite them with new equation numbers.

$$\frac{s_2}{s_1} = \left(1 - \frac{V_w}{V_1}\right)^2 \text{(Headwind)} \tag{13.6}$$

$$\frac{s_2}{s_1} = \left(1 + \frac{V_w}{V_1}\right)^2 \text{(Tailwind)} \tag{13.7}$$

HAZARDS OF HYDROPLANING

Hydroplaning is a phenomenon that occurs when a tire loses contact with the runway surface due to a build-up of water in the tire-ground contact area. NASA has researched this problem since the late 1950s and has identified three forms of hydroplaning: dynamic, viscous, and reverted rubber.

Dynamic hydroplaning is caused by the build-up of hydrodynamic pressure at the tire-pavement contact area. The pressure creates an upward force that effectively lifts the tire off the surface. When complete separation of the tire and pavement occurs, the condition is called *total dynamic hydroplaning*, and wheel rotation will stop.

Figure 13.10 shows the forces on a tire without the presence of water on the runway. Figure 13.10(a) shows a standing tire with only the forces being the weight of the aircraft and the ground reaction opposing it.

As the aircraft moves to the left, as in Figure 13.10(b), the ground friction causes a spin-up moment and wheel rotation results. When the tire is rolling freely at a fixed speed on a dry runway, the vertical ground reaction shifts forward of the axle and a spin-down moment that offers resistance to the wheel rotation is developed.

When these two moments are equal, the wheel is turning at a constant RPM.

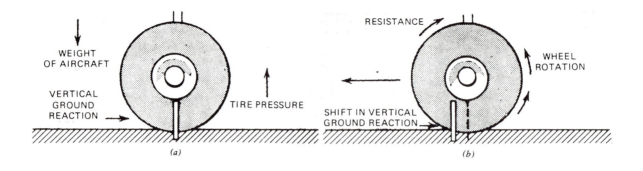

Figure 13.10 *Forces on tire: (a) static condition; (b) rolling tire.*

The introduction of water on the runway leads to dynamic hydroplaning, which is illustrated in Figure 13.11.

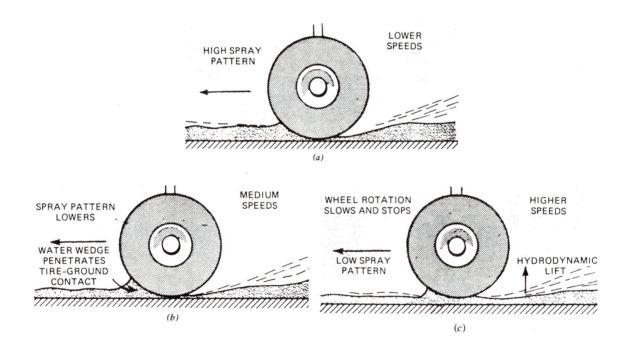

Figure 13.11 *Hydroplaning forces on tire: (a) low speed; (b) medium speed; (c) high speed.*

Deep fluid on the runway creates additional drag on the tire when it is displaced from the tire path, and a high spray pattern is produced as shown in Figure 13.11(a).

As the forward speed of the aircraft is increased (Figure 13.11(b)), the spray pattern thrown up by the tire lowers and the wedge of water penetrates the tire-ground contact area and produces a hydrodynamic lift force on the tire. This is *partial hydroplaning*. As the speed increases the spray pattern becomes flatter, and the wedge of fluid penetrates farther into the ground contact area until at some high forward speed complete separation of the tire and runway takes place and *total hydroplaning* occurs (Figure 13.11(c)).

The ground friction is progressively reduced as the wedge of water penetrates beneath the tire. It approaches zero at total hydroplaning, and the spin-down moment causes the tire to stop the wheel rotation. Obviously no braking action is available when the wheel is not making contact with the runway and has stopped rotating.

Total dynamic hydroplaning is more of a landing than a takeoff problem, however crosswind takeoffs are dangerous under these conditions. The approximate speed at which total dynamic hydroplaning occurs is:

$$V_H = 9\sqrt{P} \qquad (13.8)$$

where:
V_H = hydroplaning speed (knots)
P = tire inflation pressure (psi)

Total dynamic hydroplaning usually does not occur unless a severe rain shower is in progress. There must be a minimum water depth present on the runway to support the tire. The exact depth cannot be predicted since other factors, such as runway smoothness and tire tread, influence dynamic hydroplaning. Both smooth runway surface and smooth tread tires will induce hydroplaning with lower water depths. While the exact depth of water required for hydroplaning has not been accurately determined, a conservative estimate for an "average" runway is that water depths in excess of 0.1 in. may induce full hydroplaning.

Viscous hydroplaning is more common than dynamic hydroplaning. Viscous hydroplaning may occur at lower speeds and at lower water depths than dynamic hydroplaning. Viscous hydroplaning occurs when the pavement surface is lubricated by a thin film of water. The tire is unable to penetrate this film, and contact with the pavement is partially lost. Viscous hydroplaning often occurs on a smooth runway pavements or where rubber deposits are present, usually in the touchdown area where a thin water film can significantly reduce the coefficient of friction.

The third type of hydroplaning is known as *reverted rubber hydroplaning*. White streaks on the runway are an indication that this type of hydroplaning has occurred. Examination of the aircraft tire will show an elliptically shaped tacky or melted rubber condition. This condition occurs when the heat that is generated during a locked wheel skid reverts the rubber to its natural state. This is called *reverted rubber*.

Methods to prevent hydroplaning include transverse runway grooving, frequent removal of rubber deposits from the touchdown areas, maximum use of aerodynamic braking, and pilot education on the subject.

SYMBOLS AND UNITS

English Symbols

N	Normal force (lb)
P	Tire inflation pressure (psi)
V_H	Hydroplaning speed (knots)

Greek Symbols

γ (gamma)	Glide angle (degrees)
μ (mu)	Coefficient of friction

EQUATIONS

13.1 $\quad L = W \cos \gamma$

13.2 $\quad D = W \sin \gamma$

13.3 $\quad \tan \gamma = \dfrac{1}{(L/D)}$

13.4 $\quad \dfrac{s_2}{s_1} = \dfrac{W_2}{W_1}$

13.5 $\quad \dfrac{s_2}{s_1} = \dfrac{1}{\sigma_2}$

13.6 $\quad \dfrac{s_2}{s_1} = \left(1 - \dfrac{V_w}{V_1}\right)^2$ (Headwind)

13.7 $\quad \dfrac{s_2}{s_1} = \left(1 + \dfrac{V_w}{V_1}\right)^2$ (Tailwind)

13.8 $\quad V_H = \sqrt{P}$

Problems

1. An airplane is in trimmed flight and its velocity is constant.

    a. It has no unbalanced forces acting on it.
    b. It has no unbalanced moments acting on it.
    c. It is in a state of equilibrium.
    d. All of the above are true.

2. An airplane is in a constant velocity engine out glide. Its altitude is decreasing. It is not in a state of equilibrium.

    a. True.
    b. False.

3. At $(L/D)_{MAX}$ an airplane will:

    a. be at minimum glide angle.
    b. be achieving maximum glide distance.
    c. will have a glide ratio equal to the numerical value of $(L/D)_{MAX}$.
    d. all of the above.

4. A steep, low power approach is more dangerous for heavy airplanes than light airplanes because:

    a. recovery from high rate of descent involves a great increase in power (or thrust).
    b. the aircraft will float down the runway and possibly overshoot the runway.
    c. flaring the aircraft to decrease the rate of descent increases induced drag.
    d. both a. and c. above.

5. A low angle, high power (or thrust) approach:

    a. may be used for propeller aircraft if a short field landing is required.
    b. is dangerous for propeller aircraft if an engine fails.
    c. should be avoided at any cost for jet airplanes as high induced drag results.
    d. all of the above.

6. Aerodynamic drag is more effective than wheel braking during the first part of a landing.

    a. True.
    b. False.

7. As a general rule, when the nose of a high performance tricycle landing gear airplane can no longer be held off the runway, it is time to start applying wheel brakes.

    a. True.
    b. False.

8. In an airplane which does not have non-skid wheel brakes, it is not a good idea to apply full brake pressure because:

    a. you might cause a blow out.
    b. you get lower coefficients of friction when wheel slippage exceeds about 15 percent.
    c. both of the above.

9. In general, if maximum wheel braking of a tricycle landing gear airplane is desired, the pilot should keep the stick full back even after the nose falls through.

    a. True.
    b. False.

10. Headwinds and tailwinds affect the landing distance by the same amount as they affect the takeoff distance.

    a. True.
    b. False.

Aircraft data for problems 11-15:

    Landing speed = 100 knots (169 fpm)
    Gross weight = 8000 lb
    Average retarding force = 2000 lb.
    Sea level standard conditions.

11. Calculate the deceleration.

12. Calculate the no-wind stopping distance.

13. If the weight is increased to 13,000 lb., calculate the no-wind stopping distance.

14. If the 8000 lb. airplane lands with a 10 knot tailwind, calculate the stopping distance.

15. If operating from an airfield where the density ratio is 0.8, calculate the no-wind stopping distance.

CHAPTER FOURTEEN

MANEUVERING PERFORMANCE

Maneuvering performance can be roughly divided into two main categories: *general turning performance* and *energy maneuverability*.

General turning performance is common to all aircraft. It involves constant altitude turns and constant speed (but not constant velocity) conditions. Energy maneuvering is relevant to tactical military missions, such as air-to-air combat maneuvers and air-to-ground maneuvers.

The discussion in this chapter will not cover energy maneuverability, but will be confined to general turning flight and the flight envelope limitations on an aircraft.

GENERAL TURNING PERFORMANCE

In turning flight the airplane is **not** in a state of equilibrium, since there must be an unbalanced force to accelerate the plane into the turn. At this point let us review the subject of acceleration.

In Chapter One, *acceleration* was defined as the change in velocity per unit of time. *Velocity* was defined as a vector quantity that involves both the speed of an object and the direction of the object's motion.

Any change in the direction of a turning aircraft is therefore a change in its velocity, even though its speed remains constant. An aircraft in a turn is subjected to an unbalanced force acting toward the center of rotation.

Newton's second law states that an unbalanced force will accelerate a body in the direction of that force. This unbalanced force is called the *centripetal force*, and it produces an acceleration toward the center of rotation known as *radial acceleration*.

The forces acting on an aircraft in a coordinated, constant altitude turn are shown in Figure 14.1. In wings level, constant altitude flight, the lift equals the weight of the airplane and is opposite to it in direction.

But in turning flight, the lift is not opposite, in direction, to the weight. Only the vertical component of the lift, called *effective lift*, is available to offset the weight. Thus if constant altitude is to be maintained, the total lift must be increased until the effective lift equals weight.

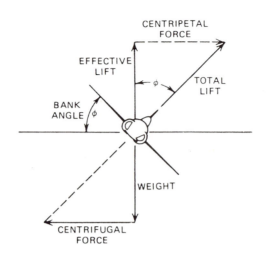

Figure 14.1 Forces on aircraft in coordinated level turn.

Solving the upper triangle in Figure 14.1:

$$\cos \phi = \frac{\text{effective lift}}{\text{total lift}} = \frac{W}{L}$$

(14.1)

The load factor G on the aircraft is defined as lift/weight

$$G = \frac{L}{W}$$

(14.2)

Inverting equation 14.1 and substituting the value of L/W into equation 14.2, we obtain:

$$G = \frac{1}{\cos \phi}$$

(14.3)

The centripetal force is found by solving the upper triangle in Figure 14.1:

$$\sin \phi = \frac{\text{centripetal force}}{\text{total lift}}$$

So

$$\text{centripetal force} = L \sin \phi$$

(14.4)

The reaction force described in Newton's third law is called the *centrifugal force* and is shown in the bottom triangle in Figure 14.1.

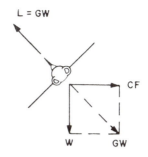

Figure 14.2   Vector diagram of forces on aircraft in turn.

A simplified vector diagram of the forces acting on an airplane in a coordinated constant altitude turn is shown in Figure 14.2. From equation 14.2 it can be seen that the total lift, L, can be replaced by its equivalent, GW.

Load Factors on Aircraft in Coordinated Turn

Equation 14.3 shows that the G's required for an aircraft to maintain altitude in a coordinated turn are determined by the bank angle alone. Type of aircraft, airspeed, or other factors have no influence on the load factor. Figure 14.3 depicts the load factors required at various bank angles.

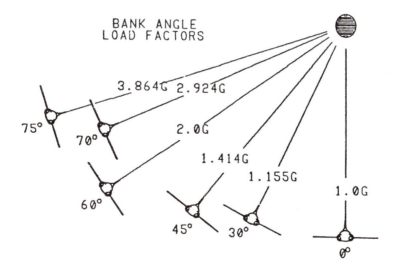

Figure 14.3   Load factors at various bank angles.

Note that the load factors required to fly at constant altitude in turning flight were determined by assuming that the wings of the aircraft provided all the lift of the aircraft. According to this analysis, the load factor at a 90° bank will be infinity and thus impossible to attain. But we have all seen aircraft fly an eight point roll, with the roll stopped at the 90° position. The secret is, of course, that lift is provided by parts of the aircraft other than the wings. If the nose of the aircraft is above the horizontal the vertical component of the thrust will act as lift.

The fuselage is also operating at some effective AOA and will also develop lift. The vertical tail and rudder will also develop lift. Figure 14.4 shows the forces on an aircraft after it has rolled 90°.

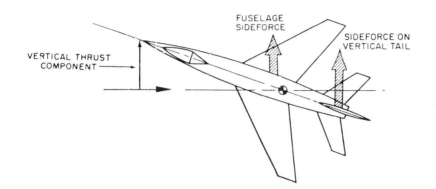

Figure 14.4 Forces on aircraft during a 90° roll.

One of the three possible limiting factors on turning performance is the structural strength limits of the airplane. We have just discussed this limitation. The bank angle determines the load factor and may, therefore, limit the turn radius.

Effect of Coordinated Banked Turn on Stall Speed

In basic aerodynamics we learned that the aircraft always develops its slow speed stall at the stall AOA. At this AOA the value of the lift coefficient is a maximum, ($C_{L(MAX)}$). So V_S occurs at $C_{L(MAX)}$.

The basic lift equation 4.1 can be rewritten as:

$$V_s = \sqrt{\frac{295L}{C_{L(MAX)}\sigma S}}$$

and since L = GW:

$$V_s = \sqrt{\frac{295GW}{C_{L(MAX)}\sigma S}}$$

It can be seen from the above equation that stall speed depends on the square root of the G loading. All other factors in the equation are constant for the same altitude.

Therefore

$$\frac{V_{s_2}}{V_{s_1}} = \sqrt{G}$$

(14.5)

where:

V_{S1} = stall speed under 1 G flight
V_{S2} = stall speed under other than 1 G flight
G = load factor for condition 2

Substituting the value of G from equation 14.3 gives:

$$\frac{V_{s_2}}{V_{s_1}} = \sqrt{\frac{1}{\cos \varnothing}}$$

(14.6)

The V-G Diagram (Flight Envelope)

Equation 14.5 shows how the stall speed increases as greater than 1 G loading is applied to an aircraft. If an aircraft is under a 4 G load, the stall speed is the square root of 4 or twice the 1 G stall speed. At zero G the stall speed is zero because no lift is being developed. This equation cannot be applied when an aircraft is developing negative load factors (square root of a negative number is impossible), but it is possible to stall the aircraft under negative G loading.

Figure 14.5 shows the first construction lines of the V-G diagram (Flight Envelope), the plot of the stall speed at various G loadings. These curved lines can also be thought of as the number of G's that can be applied to the aircraft before it will stall at any airspeed. They are called the *aerodynamic limits* of the aircraft. It is impossible to fly to the left of these curves, because the aircraft is stalled in this region.

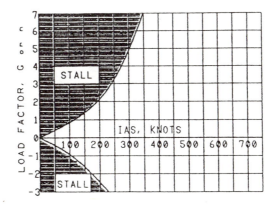

Figure 14.5   First stage construction of V-G diagram.

211

Equation 14.3 relates the G loading to the bank angle. Rearranging this equation we see that the bank angle is the angle whose cosine is $\frac{1}{G}$. Mathematically, this is:

$$\emptyset = \text{arc cos}\left(\frac{1}{G}\right) \tag{14.7}$$

Table 14.1 shows the bank angles that produce the load factors shown in Figure 14.5.

Table 14.1 Load Factors at various Bank Angles

Load Factor, G	Bank Angle, $\emptyset°$
1	0
2	60
3	70.5
4	75.5
5	78.5
6	80.4
7	81.8

The second stage of construction of the V-G diagram consists of drawing horizontal lines at the positive and negative *limit load factors*, LLFs. These limits are specified for each model aircraft and are shown in Figure 14.6 as "LIMIT LOAD FACTOR +7G" and as "LIMIT LOAD FACTOR –3G."

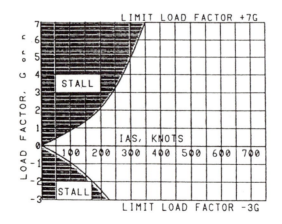

Figure 14.6   Second stage construction of V-G diagram.

The structural limit load factors show the maximum G's that may be imposed on the aircraft in flight without damaging the structure. There are several important facts about load factors that should be understood:

First, the limits are for a certain gross weight, and they change if the gross weight differs from that specified. Second, the limits are for symmetrical loading only. Third, the assumption is made that there is no corrosion, metal fatigue or other damage to the aircraft. Fourth, up-gusts and turbulent air can add to pilot imposed load factors.

The aircraft designer designs a certain strength into the airframe. The allowable total strength divided by the weight of the aircraft determines the limit load factor. If the design weight is exceeded, then the number of G's that can be accepted by the airframe must decrease. This is an inverse proportion:

$$\frac{LLF_2}{LLF_1} = \frac{W_1}{W_2}$$

(14.8)

During unsymmetrical maneuvers, such as a rolling pull-out, the wing that is rising will have more lift on it, and thus a greater load factor, than the downgoing wing. This is shown in Figure 14.7. The accelerometer is located on the center line of the aircraft and will measure the average load factor. In Figure 14.7, the pilot reads 5 G's, but the upgoing wing will actually be subjected to 7 G's.

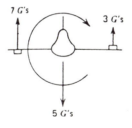

Figure 14.7 Antisymmetrical loading.

Maneuver Speed

An interesting point on the V-G diagram is the intersection of the aerodynamic limit line and the structural limit line. The aircraft's speed at this point is called the *maneuver speed*, V_p, commonly called the corner speed. At any speed below this speed the aircraft cannot be overstressed. It will stall before the limit load factor is reached. Above this speed, however, the aircraft can exceed the limit load factor before it stalls. At the maneuver airspeed the aircraft's limit load factor will be reached at the lowest possible airspeed.

The maneuver airspeed is shown in Figure 14.8.
It can be calculated by:

$$V_P = V_S \sqrt{LLF} \qquad (14.9)$$

where:

V_P = maneuver speed (knots)
V_S = stall speed (knots)
LLF = limit load factor

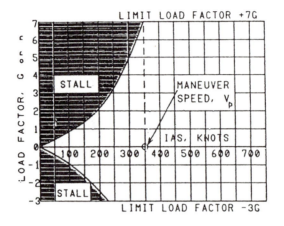

Figure 14.8   Maneuver (corner) speed.

All missions that an aircraft is designed to perform can be accomplished without exceeding the limit load factors, but on occasion a pilot will cause the aircraft to exceed these load factors. If this happens, the aircraft will probably suffer permanent objectionable deformation, resulting in costly repairs, but it will not necessarily result in failure of the primary structure.

There is another load factor that, if exceeded, may lead to catastrophic failure of the airframe. This load factor is called the *ultimate load factor*, ULF. Numerically the positive and negative ultimate load factors are 1.5 times the limit load factors. These are shown in Figure 14.9 and are labeled as "STRUCTURAL FAILURE."

The high speed limit of the V-G diagram is called the *never exceed velocity*, V_{NE}, or more commonly, the *redline speed*. It is shown in Figure 14.9. If this speed is exceeded structural failure may occur.

Stay inside of the flight envelope and you will stay out of the structural damage area.

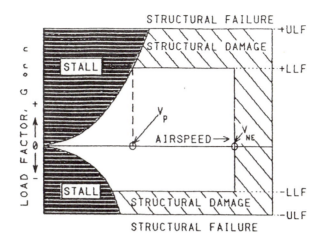

STRUCTURAL FAILURE

STRUCTURAL DAMAGE

AIRSPEED →

STRUCTURAL DAMAGE

STRUCTURAL FAILURE

Figure 14.9   Ultimate load factors.

Radius of Turn

A body traveling in a circular path is undergoing an acceleration toward the center of rotation. This is called *radial acceleration*, a_R. It is a function of the velocity of body, V, and the radius, r, of the circle:

$$a_R = \frac{V^2}{r} \tag{14.10}$$

As was shown in Figure 14.1, the horizontal component of the total lift is the centripetal force that causes the radial acceleration. Also, the reaction force to the centripetal force, called the *centrifugal force*, (CF), is equal in magnitude and opposite in direction to the centripetal force.

The centripetal force in an automobile, which is making a turn, is generated by the friction of the car's tires on the pavement. The force that pulls the driver outward, away from the center of the turn, is the centrifugal force. Since the centrifugal force is equal to the centripetal force in magnitude and results from the radial acceleration, it is, according to Newton's second law, equal to the mass times the radial acceleration:

$$CF = ma_R = \left(\frac{W}{g}\right)\left(\frac{V^2}{r}\right) \tag{14.11}$$

Equation 14.4 showed that this force is:

$$CF = L \sin \o$$

Equating CF values gives:

$$L \sin \o = \frac{WV^2}{gr} \tag{14.12}$$

215

Equation 14.1 then can be rewritten as:

$$L \cos \varnothing = W \tag{14.13}$$

By dividing equation 14.12 by equation 14.13 we obtain:

$$\tan \varnothing = \frac{V^2}{gr} \text{ or } r = \frac{V^2}{g \tan \varnothing} \tag{14.14}$$

This is the *radius of turn* equation.

Equation 14.14 was derived in basic units, and thus the velocity is in units of feet per second. If V is measured in knots:

$$r = \frac{V_K^2}{11.26 \tan \varnothing} \tag{14.15}$$

where
r = radius of turn (ft)
V_K = velocity (knots TAS)
$\varnothing$ = bank angle (degrees)

Rate of Turn

The *rate of turn*, ROT, is primarily used during instrument flight and is the change in heading of the aircraft per unit of time:

$$ROT = \frac{g \tan \varnothing}{V_{FPS}} \text{ (radians/sec)}$$

If velocity is in knots and ROT is measured in degrees per second:

$$ROT = \frac{1091 \tan \varnothing}{V_K} \tag{14.16}$$

The importance of bank angle and velocity can be seen in equations 14.15 and 14.16. High bank angles and slower airspeeds produce small turn radii and high ROT.

The *maneuver speed* is the speed at which highest bank angle can be achieved at minimum airspeed. Hence minimum turn radius and maximum rate of turn will be realized at this speed.

Figure 14.10 is a chart that shows turn radii and rates of turn for various bank angles and airspeeds for an aircraft making a coordinated, constant altitude turn.

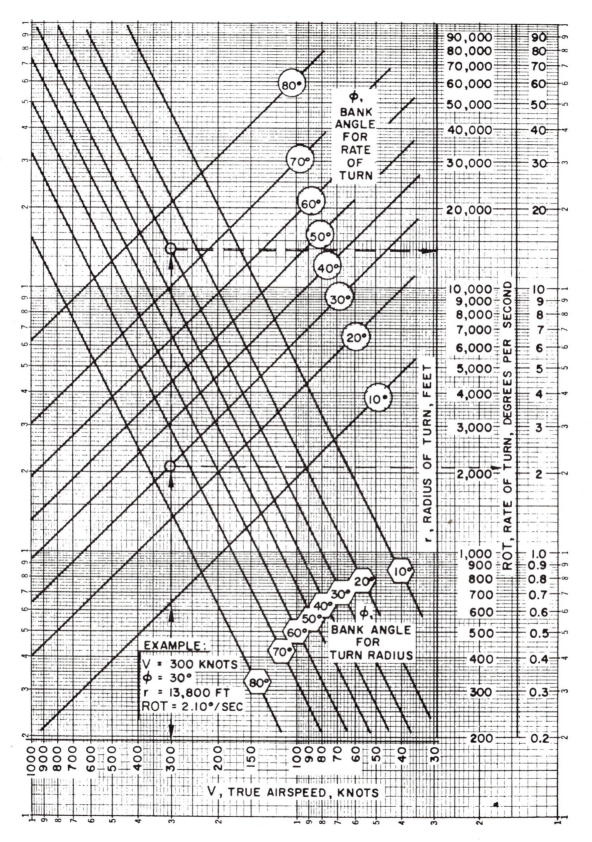

Figure 14.10   Constant Altitude turn Performance.

Turn Radius Limited by Aerodynamic and Structural Limits

A curve of the turn radius at various airspeeds is shown in Figure 14.11. The aerodynamic limits are so labeled on the curve at the left side of the figure. This curve is drawn as follows: From the V-G diagram of the aircraft, pick any airspeed between the stall speed and the maneuver airspeed, read the load factor that occurs at the aerodynamic limit line, and calculate the bank angle that produces this load factor in constant altitude turning flight (see Table 14.1). Enter Figure 14.10 with these values, or use equation 14.15, and determine the turn radius. Repeat for other airspeeds, plot and draw the curve.

The structural limit curve on the right of Figure 14.11 is determined by the limit load factor of the aircraft. This limit determines the amount of bank angle that the aircraft can structurally withstand. Once this is found, the turn radius is a function of the airspeed.

Substituting various airspeeds yields corresponding values of turn radius. The intersection of the two curves in Figure 14.11 shows the minimum turn radius and the airspeed at which it occurs.

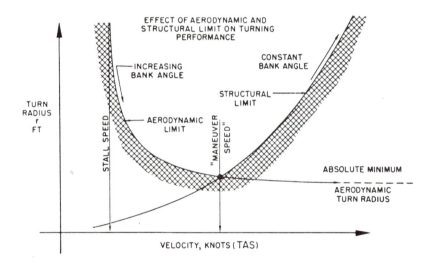

Figure 14.11   Turn radius versus airspeed.

Turn Radius Limited by Thrust Available

We know that in making minimum radius turns at constant altitude the load factor increases as 1/cos ø (equation 14.3). The increased load factor has a strong effect on the induced drag of the airplane.

In our discussion of induced drag in Chapter Five, we saw that the rearward component of the total lift vector was induced drag and that the vertical component was the effective lift. These are shown in Figure 14.12.

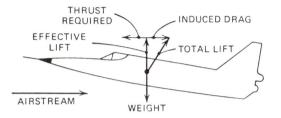

Figure 14.12   Forces on complete aircraft.

We see that the effective lift must equal the weight of the plane, if constant altitude is to be maintained. We also see that more thrust is required to overcome the increase in induced drag, if constant airspeed is to be maintained. Thus it may be that the turn radius may be limited by thrust available instead of structural considerations.

Figure 14.13 shows a thrust limited aircraft's turn radius plotted against true airspeed. The aerodynamic limits are the same as were described for Figure 14.11. The thrust limit curve is determined by a combination of bank angle and airspeed.

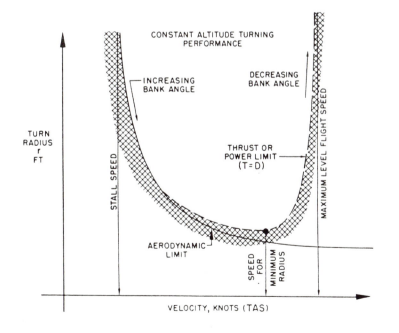

Figure 14.13   Thrust limited turn radius.

SYMBOLS AND UNITS

a_R	Radial acceleration (fps^2)
arc cos	"The angle whose cosine is"
LLF	Limit load factor (G units)
r	Radius of turn (ft)
ROT	Rate of turn (degrees per sec)
ULF	Ultimate load factor
V_{NE}	Never exceed speed (knots)
V_P	Maneuver speed (knots)

EQUATIONS

14.1 $\cos \o = \dfrac{W}{L}$

14.2 $G = \dfrac{L}{W}$

14.3 $G = \dfrac{1}{\cos \o}$

14.4 $CF = L \sin \o$

14.5 $\dfrac{V_{s2}}{V_{s1}} = \sqrt{G}$

14.6 $\dfrac{V_{s2}}{V_{s1}} = \sqrt{\dfrac{1}{\cos \o}}$

14.7 $\o = \text{arc cos}\left(\dfrac{1}{G}\right)$

14.8 $\dfrac{LLF_2}{LLF_1} = \dfrac{W_1}{W_2}$

14.9 $V_P = V_s\sqrt{LLF}$

14.10 $a_R = \dfrac{V^2}{r}$

14.11 $CF = \dfrac{WV^2}{gr}$

14.12 $L \sin \o = \dfrac{WV^2}{gr}$

14.13 $L \cos \o = W$

14.14 $r = \dfrac{V^2}{g \tan \o}$

14.15 $r = \dfrac{V_K{}^2}{11.26 \tan \o}$

14.16 $ROT = \dfrac{1091 \tan \o}{V_K}$

PROBLEMS

1. In a constant altitude, constant airspeed turn, an an aircraft is in a state of equilibrium.

    a. True.
    b. False.

2. The "G" forces on an airplane in a constant altitude turn depend on:

    a. airspeed.
    b. type of airplane (jet or prop).
    c. bank angle.
    d. all of the above.

3. An airplane performing an 8 point roll can stop in the 90° position because:

    a. the wings are not supporting the weight.
    b. the fuselage and vertical tail are supporting much of the weight.
    c. the pilot pulls the nose above the horizontal thus some of the thrust helps support the weight.
    d. all of the above.

4. When an airplane is in a constant altitude bank the stall speed:

    a. remains the same as in level flight.
    b. increases as the square root of the "G"s.
    c. increases as the square root of $1/\cos \phi$.
    d. both b. and c. above.

5. What can we find from the "Aerodynamic Limit" line on a V-G diagram?

    a. It shows maximum "G"s that can be pulled at your airspeed.
    b. It shows stall speed when "G"s are being pulled.
    c. It is impossible to fly to the left of this line as the airplane will stall there.
    d. All of the above.

6. Limit Load Factor (LLF) is also called "Maximum Design G".

    a. True.
    b. False.

7. If the weight of an airplane (W_2) is increased above the design gross weight (W_1), the limit load factor:

    a. remains the same value.
    b. increases by the weight ratio, W_2/W_1.
    c. decreases by the weight ratio, W_2/W_1.
    d. none of the above.

8. If an airplane is flying in symmetrical flight at its maneuver speed:

    a. it can not be overstressed.
    b. it can make the smallest turn radius.
    c. it can make the highest rate of turn.
    d. all of the above.

9. The "Constant Altitude Turn Performance Chart" (Figure 14.10):

    a. can be used for only one particular model of airplane.
    b. does not show if the airplane is overstressed.
    c. is good for any model airplane.
    d. both b. and c. above.

10. When an airplane makes a level banked turn, the airspeed slows (unless throttle is added). This is because:

    a. Parasite drag is increased the most.
    b. Induced drag is increased the most.
    c. Profile drag is increased the most.

11. A twin engine airplane has an engine failure shortly after takeoff. The pilot tries to turn back toward the field by making a left hand turn but crashes after making a 180° turn. The wreckage diagram shows that the turn radius was 6000 ft. The airspeed during the turn is estimated to be 200 knots TAS. Calculate the bank angle of the airplane.

12. Using Figure 14.10 verify the answer to problem 11.

13. Assuming a level turn, find the G's on the airplane in problem 11 during the turn.

14. For the airplane in problem 11, calculate the rate of turn and the time to make the 180° turn.

15. Using Figure 14.10 verify the ROT you calculated in problem 14.

CHAPTER FIFTEEN

LONGITUDINAL STABILITY AND CONTROL

In addition to adequate performance, an aircraft must have satisfactory handling qualities. An aircraft must be able to maintain uniform flight and be able to recover from the effects of disturbing influences, such as gusts.

This ability is called the *stability* of an aircraft. Adequate stability is necessary to minimize the workload of the pilot. In some cases, such as helicopter flight, it is necessary to provide artificial stabilization by use of *automatic stabilization equipment*, (ASE).

Control is the response of an aircraft to the directions of a pilot. For an aircraft to respond to the controls, its stability must be overcome. Stability and control are often compared to a seesaw, with stability at one end and control at the other. The more stability an aircraft has, the less controllability it has, and vice versa.

Modern complex, high performance aircraft have stability problems that are very complicated and are beyond the scope of this book. Our discussion here is, therefore, basic and simple. Some simplifying assumptions are made to keep the discussion from getting too involved.

DEFINITIONS

Equilibrium. An aircraft is said to be in a state of equilibrium if the sum of all moments and forces at its center of gravity are equal to zero. This means that there is no pitching, yawing, or rolling, nor is there any change of velocity taking place.

Static Stability. Static stability is the **initial tendency** of an aircraft to move, once it has been displaced from its equilibrium position. If it has the tendency to return to its equilibrium position, it is said to have *positive static stability*.

This is illustrated by the ball in Figure 15.1(a). Point A is the equilibrium position of the ball. If the ball is moved to point B, it has the tendency to return toward point A. It has positive static stability. Note that for the ball to have positive static stability it is not important that the ball actually returns to point A, only that it has the tendency to return.

If an aircraft that has been disturbed from its equilibrium position has the initial tendency to move farther away from its equilibrium position, it is said to have *negative static stability*.

This is illustrated by the ball in Figure 15.1(b). The ball has been displaced from its equilibrium position at point A to point B. The initial tendency of the ball is to move farther away from point A. The ball has negative static stability.

If an aircraft is disturbed from its equilibrium position and has the tendency neither to return nor to move farther away from its equilibrium position, it is said to have *neutral static stability*. This is illustrated by the ball on a flat surface, as shown in Figure 15.1(c). If the ball is displaced to point B, it does not have a tendency to return to point A, nor does it have a tendency to move away from point A. This is neutral static stability.

For an aircraft to have positive stability, it must **first** have positive static stability.

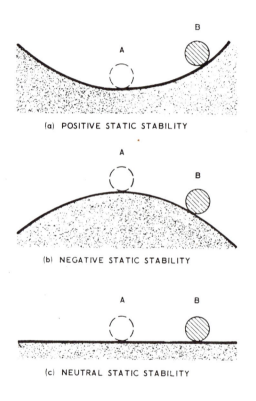

(a) POSITIVE STATIC STABILITY

(b) NEGATIVE STATIC STABILITY

(c) NEUTRAL STATIC STABILITY

Figure 15.1   Static stability.

Dynamic Stability. Dynamic stability is the movement of an aircraft with respect to time. If an aircraft has been disturbed from its equilibrium position and the maximum displacement decreases with time, it is said to have *positive dynamic stability*.
If the maximum displacement increases with time, it is said to have *negative dynamic stability*. If the displacement remains constant with time, it is said to have *neutral dynamic stability*.

OSCILLATORY MOTION

Aircraft motions are oscillatory in nature. Therefore, let us consider the possibilities of these oscillations. As we just mentioned, an aircraft must have positive static stability, but this does not ensure that it will have positive dynamic stability as well. Consider the graph of aircraft displacement versus time shown in Figure 15.2.

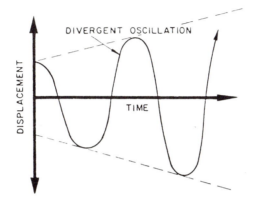

Figure 15.2   Positive static and negative dynamic stability.

The above graph shows that the first reaction to the displacement is to return toward the equilibrium position. This is positive static stability. The subsequent oscillations, however, are divergent, and so the dynamic stability is negative. This is a case of positive static and negative dynamic stability, and it is unacceptable.

Another possibility is that of positive static stability and neutral dynamic stability. This is shown in Figure 15.3.

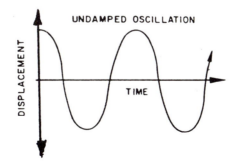

Figure 15.3   Positive static and neutral dynamic stability.

No damping of the oscillations occurs in the above graph. This is also unacceptable.

The desired combination of positive static and positive dynamic stability is called a *damped oscillation.* An airplane will return to its equilibrium condition when this occurs.

This is shown in Figure 15.4.

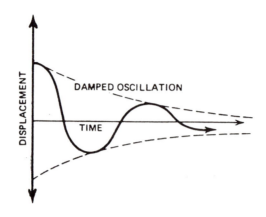

Figure 15.4   Positive static and positive dynamic stability.

AIRPLANE REFERENCE AXES

To better envision the forces and moments acting on an airplane, we assign three mutually perpendicular reference axes, all intersecting at the CG. These are shown in Figure 15.5.

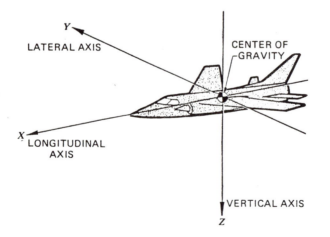

Figure 15.5   Airplane reference axes.

The longitudinal axis is assigned the letter X for identification, the lateral axis is called the Y axis, and the vertical axis is the Z axis. The positive direction of the axes is determined by positioning the thumb, index finger, and middle finger of the right hand so that they are at right angles to each other. Point the thumb in the forward direction of the X axis, and the other fingers point to the positive directions of the Y and Z axes.

There are three moments possible about the three axes. These are given three identifying letters that occur in sequence in the alphabet. The rolling moment is named L (to avoid confusion with lift we will call this L'), the pitching moment is M, and the yawing moment is N. The positive direction of these moments are determined by using the right hand rule, as illustrated in Figure 15.6.

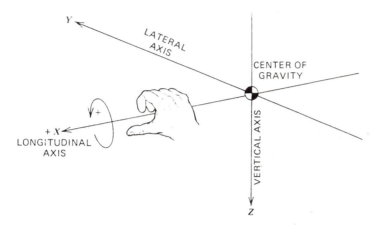

Figure 15.6   *Establishing positive moment direction.*

The right thumb points to the positive axis, and the curvature of the fingers shows the direction of the positive moments.

The completed axes and moment illustration is shown in Figure 15.7.

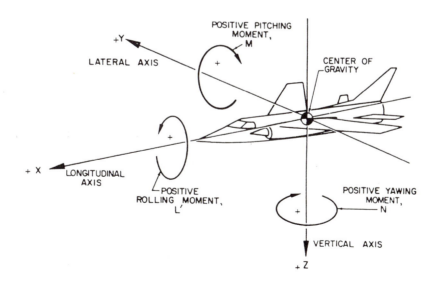

Figure 15.7   *Airplane axes and moment directions.*

Longitudinal stability and control refer to the behavior of the airplane in pitch, that is, movements of the longitudinal axis about the lateral axis. This movement is illustrated in Figure 15.8. In the pure pitching case, no rolling or yawing of the aircraft takes place.

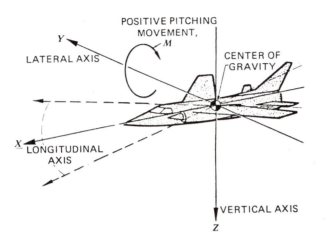

Figure 15.8   Movement of the longitudinal axis in pitch.

STATIC LONGITUDINAL STABILITY

An airplane is said to have *positive static longitudinal stability* if it tends to maintain a constant AOA in flight, once it has been trimmed to that angle. If an airplane is trimmed so that it has zero pitching moments at some AOA and is in a state of equilibrium, and is then disturbed in pitch and tends to return to the trimmed AOA, it is said to have positive static longitudinal stability. An airplane which has negative static longitudinal stability, on the other hand, will continue to pitch away from the trimmed AOA. If an airplane has neutral static longitudinal stability it will remain at whatever AOA the disturbance has caused.

It is very important for an airplane to have positive static longitudinal stability. A stable airplane is easy and safe to fly. It can be trimmed at any desired speed and will tend to maintain that speed. An airplane which has negative static longitudinal stability, on the other hand, will be practically impossible to trim and will require the pilot's undivided attention at all times.

The Pitching Moment Equation

The pitching moment about the aircraft CG is:

$$M_{(CG)} = C_{M(CG)}qSc \qquad (15.1)$$

where:

$M_{(CG)}$ = pitching moment about the CG (ft-lb)
$C_{M(CG)}$ = coefficient of pitching moment about CG
q = dynamic pressure (psf)
S = wing area (ft^2)
c = mean aerodynamic chord, MAC (ft)

Rearranging equation 15.1:

$$C_{M(CG)} = \frac{M_{CG}}{qSc} \tag{15.2}$$

Because the values of q, S, and c are always positive, it follows that for a nose up, (+), pitching moment, the value of $C_{M(CG)}$ must also be positive; and for a nose down, (–), pitching moment, the value of $C_{M(CG)}$ must be negative.

Graphic Representation of Static Longitudinal Stability

A plot of the variation of $C_{M(CG)}$ at different values of C_L, (different AOAs), for an airplane with positive static longitudinal stability is shown in Figure 15.9.

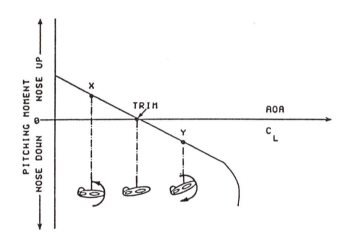

Figure 15.9   Positive static longitudinal stability.

The trim point is the value of C_L where the aircraft has no pitching moment. At all values of C_L above the trim point, such as point y, the aircraft will have a nose down, (–), pitching moment. If the aircraft is disturbed by an up-gust, the AOA (and C_L) will be increased.

For stability, a nose-down pitching moment is required, so this is a stabilizing condition.

If the airplane is disturbed by a down-gust, the AOA and value of C_L will be reduced. This is represented by point x in Figure 15.9. The value of $C_{M(CG)}$ at point x is positive, and a nose-up pitching moment results.

This is what is required for static stability. Thus a negative slope on this graph represents an aircraft with positive static stability.

Conversely, a positive slope would indicate an unstable aircraft, and a zero slope would represent an aircraft with neutral stability. These are shown in Figure 15.10.

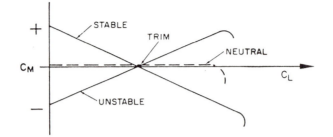

Figure 15.10   Types of static longitudinal stability.

There are different degrees of stability. Some aircraft tend to return to their equilibrium positions faster than others. Again, using the analogy of the ball in the curved container, the degree of stability can be illustrated as shown in Figure 15.11.

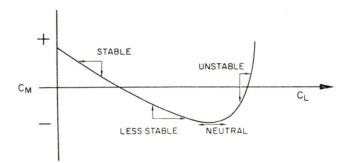

Figure 15.11   Degrees of positive static stability.

Degrees of stability are shown on the $C_{M(CG)}$– C_L plot by the slope of the curve. Steeper slope of the stable curve shows more stability, and steeper slope of the unstable curve shows more instability.

An airplane can be stable at lower angles of attack but may be unstable at higher angles of attack. This would indicate a pitch-up problem at high AOA. This is shown in Figure 15.12.

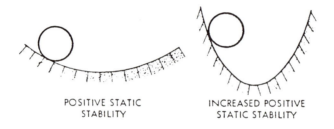

Figure 15.12   Aircraft static longitudinal stability.

Contribution of Aircraft Components to Pitch Stability

Wings

The static stability contribution of the wings depends on the relative position of the aerodynamic center AC and the center of gravity CG of the airplane.

Consider a tailless (flying wing) airplane with a symmetrical airfoil section as shown in Figure 15.13. We will consider two possibilities of CG and AC location:

(1) where the CG is located forward of the AC as shown in Figure 15.13(a) and

(2) where the CG is located behind the AC as shown in Figure 15.13(b).

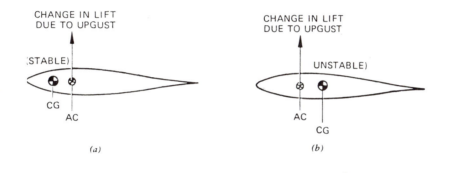

Figure 15.13   Effect of CG and AC location on static longitudinal stability.

If the airplane in Figure 15.13(a) experiences an up-gust, the AOA will be increased, and the lift at the AC will increase. The airplane will then rotate about the CG, and this nose-down moment will tend to rotate the airplane and return it to its equilibrium AOA. This is the stable condition.

Should the airplane in Figure 15.13(b) experience an up-gust, the increase in lift at the AC will rotate the airplane about its CG and create a nose-up pitching moment and rotate the airplane away from its equilibrium position. This is the unstable condition. The CG must be ahead of the AC for a stable flying wing.

The airplane in Figure 15.13(a) is stable in pitch, but it is not in equilibrium. The conditions for equilibrium require that there be no unbalanced forces or unbalanced moments acting on the airplane.

The first requirement is easily satisfied by adjusting the AOA so that lift equals weight. The second requirement, however, has not been met. The forces of lift and weight create a nose-down pitching moment, which must be canceled out by an opposing nose-up pitching moment.

In our discussion of pressure distribution on airfoils in Chapter Three, we saw that the forces acting on the top and bottom surfaces of a symmetrical airfoil were located in the same position along the chord and that no pitching moment resulted from their location (see Figure 3.9).

The pressure location on cambered airfoils, on the other hand, was found not to be located at the same chordwise position, and so a nose-down pitching moment resulted from positive camber (Figure 3.10). A negatively cambered airfoil produces a nose-up pitching moment, which is what we need to cancel the nose-down moment caused by the lift and weight vectors. This is shown in Figure 15.14. Delta wing airplanes use reflexed (negatively cambered) trailing edges to create the nose-up pitching moments required for equilibrium.

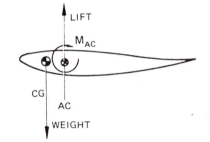

Figure 15.14   Static longitudinally stable flying wing in equilibrium.

The overall contribution of the wings of an aircraft to the static longitudinal stability depends on the location of the AC and the CG of the aircraft. If the AC is behind the CG, the contribution is stabilizing. If the AC is ahead of the CG, the contribution is destabilizing. The trend in recent years is to locate the wings farther back on the fuselage, thus increasing the aircraft's pitch stability. There is, however, one disadvantage in doing this. Consider the vertical forces acting on the aircraft if the AC is located behind the CG as shown in Figure 15.15.

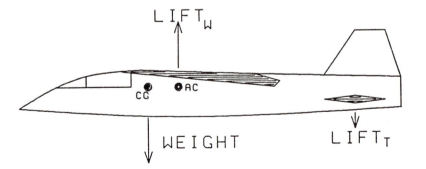

Figure 15.15   Airplane with static longitudinal stability.

It is assumed that this airplane has a symmetrical airfoil so it has no pitching moment due to the pressure distribution on the airfoil. For balance, the airplane in Figure 15.15 must have a download on the tail, ($LIFT_t$). The lift on the wing, ($LIFT_w$), must equal the weight of the aircraft plus the lift on the tail. Thus it must operate at a higher AOA than if the lift on the tail acted upward. Higher angles of attack produce more drag. One of the newer concepts is to reduce the static stability by moving the CG backward, and thus reduce the drag. This may require automatic stabilization devices.

Fuselage

A streamlined fuselage has a pressure distribution similar to that of a body of revolution when placed in an airstream. The pressure distribution about such a body is shown in Figure 15.16. It can be seen that there is no net lift developed by the pressure distribution in Figure 15.16, but a nose up pitching moment is developed by an up-gust, so the pitching moment is destabilizing. Thus, the fuselage is a destabilizing component.

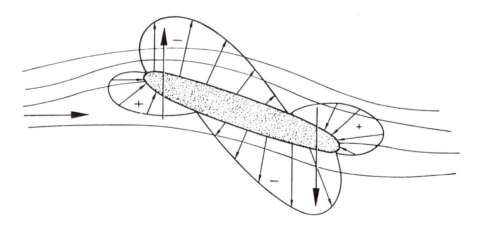

Figure 15.16   Pressure distribution about a body of revolution.

Engine Nacelles

The direction of the airflow through a propeller disc or through a turbojet engine is not changed if the axis of the engine is in line with the aircraft's flight path. If the engine axis is at an angle with the relative wind, however, the airflow is turned so that it flows in the direction of the engine axis. When this happens, a side force is developed on the propeller shaft or on the side of the jet engine intake, in accordance with Newton's third law.

This is illustrated by the aircraft shown in Figure 15.17.

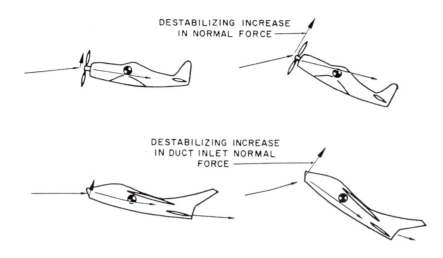

DESTABILIZING INCREASE
IN NORMAL FORCE

DESTABILIZING INCREASE
IN DUCT INLET NORMAL
FORCE

Figure 15.17   Engine nacelle location contribution to pitch stability.

The axes of the above aircraft are at a positive AOA to the relative wind, and the resulting force creates a nose-up pitching moment. This is a destabilizing moment If the engines are mounted so that the propellers or jet intakes are behind the CG and the axis of the engine makes a positive angle with the relative wind, the resulting upward side force produces a nose-down moment. This is a stabilizing moment.

Propellers or jet engine intakes located forward of the CG are destabilizing components, and propellers or jet engine intakes located behind the CG are stabilizing components.

Horizontal Stabilizer

As the name implies, the horizontal stabilizer is a strongly stabilizing influence on static longitudinal stability. It is usually a symmetrical airfoil because it must produce both upward and downward airloads. The contribution of the horizontal stabilizer to the pitch stability of the aircraft can be seen in Figure 15.18. Figure 15.18(a) shows that if an up-gust causes the aircraft to pitch up, then an upward lift is developed by the horizontal tail. This creates a nose-down moment, which is stabilizing. In Figure 15.18(b) the opposite effect is achieved when the aircraft is pitched downward by a down-gust. A nose-up moment is developed in this case.

The degree of stability produced by the tail is determined by the tail size and the moment arm to the aircraft's CG. The tail area (ft^2) multiplied by the moment arm (ft) equals the dimension of ft^3, often called the "tail volume," and it is an indication of the stabilizing effectiveness of the horizontal stabilizer.

An increase in either the size of the surface, or the distance between the CG of the airplane and the AC of the stabilizer will increase the tail volume and thus the stability of the aircraft.

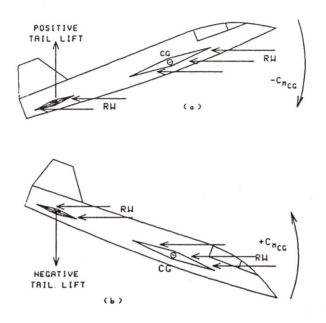

Figure 15.18   Lift of horizontal stabilizer produces a stabilizing moment.

Total Airplane

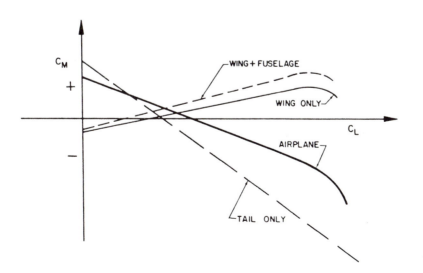

Figure 15.19   Typical build-up of aircraft components.

Figure 15.19 shows a typical build-up of wing with its AC forward of the CG, the wing plus the fuselage, the horizontal tail alone, and the total airplane. The horizontal stabilizer has enough stability to overcome the negative stability of the wing and fuselage combined. Stability of the engine location is not shown in this figure.

Effect of CG Position

Varying the CG position has a great effect on the longitudinal static stability of an airplane as shown in Figure 15.20. As was explained earlier and illustrated in Figure 15.13, forward positioning of the CG results in increased pitch stability. This is shown by the slope of the C_M-C_L curves in Figure 15.20. At 40 percent MAC the slope is zero, and the aircraft has neutral stability at this point. Moving the CG behind this point results in an unstable condition.

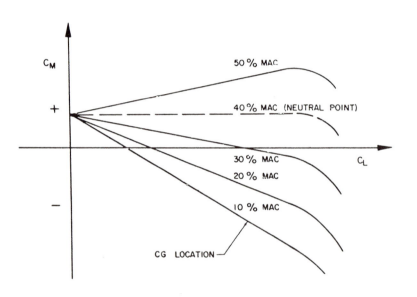

Figure 15.20   Effect of CG location on static longitudinal stability.

Stick Fixed-Stick Free Stability

If the elevators (or other control surfaces) are allowed to float free, they will be moved from their neutral positions by outside disturbances, and the total stabilizing surface will effectively be reduced.

The stability of an aircraft with "free floating" control surfaces is known as *stick free stability*. It will be less stable than an aircraft with the controls held in a fixed position, known as *stick fixed stability*.

Figure 15.21 shows the C_M-C_L curves for both conditions. The slope of the stick fixed condition is greater; hence, the stability is greater for the stick fixed case. Unless the airplane is equipped with irreversible powered controls, better stability will be realized if the controls are held in the neutral position.

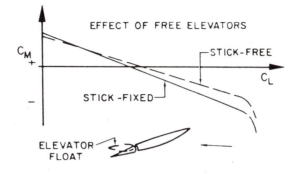

Figure 15.21   Stick free-stick fixed stability.

DYNAMIC LONGITUDINAL STABILITY

Stability considerations that we have discussed up to now have dealt with static stability. Dynamic stability involves the response of an aircraft to disturbances over a period of time. Dynamic stability exists when the amplitude of these disturbances dampens out with time. The reaction of an aircraft to disturbances differs, depending on whether the controls are in the "stick free" or the "stick fixed" configuration.

First, let us consider the "stick free" or reversible controls aircraft. Most light general aviation aircraft are in this category. If a control surface is capable of being manually moved from outside the aircraft, it is in this category.

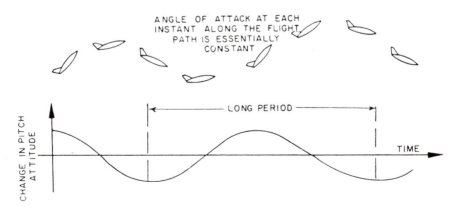

Figure 15.22   Phugoid longitudinal dynamic mode.

There are two types of dynamic oscillations possible. One is the long period with poor damping oscillation, called the *phugoid* mode. This is shown in Figure 15.22.

The phugoid oscillation is one where the airspeed, pitch, and altitude of the airplane vary widely, but the AOA remains nearly constant. The motion is so slow that the effects of inertia forces and damping forces are very low. The whole phugoid can be thought of as a slow interchange between kinetic and potential energy.

The second mode is a short period mode as shown in Figure 15.23.

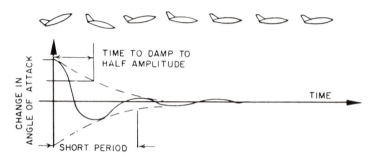

Figure 15.23   Short period dynamic mode.

The short period mode in the stick free configuration is caused by the elevator flapping about its hinge line. A typical flapping mode may have a period of 0.3 to 1.5 seconds and has heavy damping of the oscillations. Recovery of this type of pitching can be accomplished by releasing the controls or, more rapidly, by holding the controls in the neutral positions. This later recovery has the effect of putting the airplane into the stick fixed configuration.

Caution must be observed not to try and dampen out the oscillations. Pilot reaction time is close to the natural period of the oscillations and inadvertent reinforcement of the pitching may result. While this may only result in a rough ride for a relatively slow light airplane it can prove disastrous for a high performance jet aircraft. If this reinforcement occurs in a high performance jet, it is called a *pilot induced oscillation,* (PIO), and it can destroy the aircraft in a matter of a few seconds.

A somewhat similar dynamic stability problem, called collective bounce, exists in helicopter flight. If an up-gust is encountered and the helicopter is forced upward, inertia forces the pilot downward in his seat. If the pilot has hold of the collective stick, this will also be forced downward, and the AOA of all of the rotors will be decreased, thus causing the helicopter to descend. Again the inertia forces cause the pilot to be forced upward, the collective stick is raised, and the process is repeated.

LONGITUDINAL CONTROL

To control an airplane it is necessary to overcome its stability. As we have seen, the farther forward the CG of the airplane is, the more static pitch stability the airplane has. However, since stability and control oppose each other, the forward CG results in lower controllability. One of the pitch control requirements is that sufficient elevator forces are available to overcome the stability of forward CG location.

Takeoff and landing maneuvers are critical as far as control forces are concerned. During takeoff an aircraft must be able to rotate to takeoff attitude. The requirement here is that the aircraft be able to attain takeoff attitude at $0.9\ V_S$.

Unlike the airborne case, the aircraft rotates about its main wheels, rather than its CG, during takeoff. The forces on the aircraft are shown in Figure 15.24.

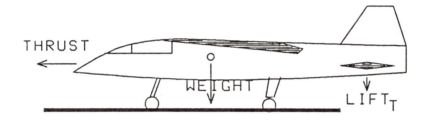

Figure 15.24 Forces producing moments during takeoff.

In many aircraft little or no lift is developed until the aircraft is rotated to takeoff attitude. The angle of incidence of the wing is selected to produce minimum drag during the takeoff run rather than to produce lift as the aircraft accelerates.

The nose-down moments that the elevator must overcome to rotate the aircraft are the moment caused by the thrust line being above the main wheels, the moment caused by the CG being ahead of the main wheels, and the nose-down moment of either a cambered wing or takeoff flaps (if used).

Another effect that is not shown on the drawing is the nose-down pitching moment caused by ground effect, which reduces the downwash over the horizontal tail of a low tailed airplane.

The elevator must be able to produce a nose-up pitching moment that will overcome all the nose-down pitching moments mentioned above.

Landing requirements include flaring the aircraft prior to touchdown and overcoming the nose-down moment caused by the airplane entering ground effect. If the airplane fulfills the takeoff requirements satisfactorily, it will usually have enough elevator control for landings.

SYMBOLS AND UNITS

$C_{M(CG)}$	Pitching moment coefficient (dimensionless)
L'	Rolling moment (ft-lb)
M_{CG}	Pitching moment about CG
N	Yawing moment
PIO	Pilot induced oscillation
X	Longitudinal axis of aircraft
Y	Lateral axis of aircraft
Z	Vertical axis of aircraft

EQUATIONS

15.1 $M_{CG} = C_{M(CG)}qSc$

15.2 $C_{M(CG)} = \dfrac{M_{CG}}{qSc}$

PROBLEMS

1. The more stability an airplane has, the easier it is to control.

   a. True.
   b. False.

2. Static stability of an airplane is:

   a. the ability to return to its equilibrium position once it has been disturbed.
   b. the long time reaction to a disturbance.
   c. the initial tendency to move toward the equilibrium position after a disturbance.
   d. none of the above.

3. Dynamic stability of an airplane is _____ to a disturbance.

   a. the immediate reaction.
   b. the long time reaction.
   c. oscillatory in nature.
   d. both b. and c. above.

4. A damped oscillation is one that:

    a. shows positive static stability.
    b. shows positive dynamic stability.
    c. shows both of the above.
    d. shows neither a. nor b. above.

5. A ball in a flower vase as compared to a ball in a soup bowl, illustrates:

    a. increased dynamic stability.
    b. decreased dynamic stability.
    c. increased static and dynamic stability.
    d. decreased static and dynamic stability.

6. Weight and balance of airplanes is important because:

    a. the CG affects the stability and control.
    b. if the CG is moved forward the static pitch stability is increased.
    c. if the CG is too far forward, the plane may not respond to elevator commands.
    d. all of the above.

7. The farther back the wings are located, the more static pitch stability an airplane has.

    a. True.
    b. False.

8. Placing the engines at the rear (DC-9):

    a. increases the static pitch stability.
    b. decreases the static pitch stability.
    c. increases the pitch control.
    d. decreases the pitch control.
    e. both a. and d. above.

9. The Phugoid oscillation is dangerous because pilot reaction can lead to PIO.

    a. True.
    b. False.

10. The most critical conditions for pitch control is that the elevators must be able to:

    a. rotate the airplane for takeoff.
    b. flare the airplane for landing.
    c. overcome a forward CG location.
    d. all of the above.

CHAPTER SIXTEEN

DIRECTIONAL AND LATERAL
STABILITY AND CONTROL

DIRECTIONAL STABILITY AND CONTROL

Directional stability and control refers to the behavior of the airplane in yaw, that is, movements of the longitudinal axis when it is rotated about the vertical axis. This rotation is caused by yawing moments and is illustrated in Figure 16.1. In the pure yawing case, there is no pitching or rolling of the aircraft.

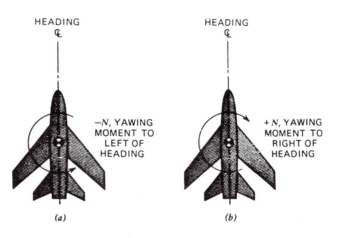

Figure 16.1 (a) Negative yawing moment causes yaw to left;
(b) Positive yawing moment causes yaw to right.

In this section we will discuss static directional stability and directional control. Dynamic directional stability is, however, coupled with dynamic roll stability, and these coupled effects are discussed later in the chapter.

Sideslip angle, (beta), β is the angle between the relative wind and the airplane's longitudinal axis. When the relative wind is from the right, the sideslip is positive in value. Figure 16.2 shows a positive sideslip angle.

STATIC DIRECTIONAL STABILITY

An airplane is said to have *positive directional stability* if it is trimmed for non-sideslip flight, is then subjected to a sideslip condition, and reacts by turning into the new relative wind so that the sideslip angle is reduced to zero.

On the other hand, if the airplane reacts to the sideslip by turning away from the new relative wind, thus increasing the sideslip angle, the airplane has *negative directional stability*. If no reaction results from sideslip, the airplane is said to have *neutral directional stability*.

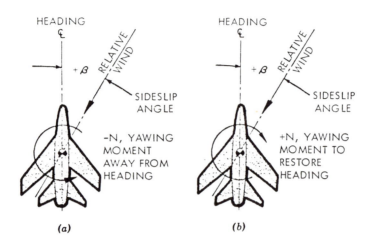

Figure 16.2   (a) Unstable; (b) stable in yaw.

Figure 16.2(a) illustrates an airplane with negative static directional stability, while Figure 16.2(b) shows an airplane with positive static directional stability. With negative static directional stability, an airplane tends to turn away from the relative wind. This leads to sideward flight that is often called "swapping ends", a clearly unacceptable condition.

The Yawing Moment Equation

The yawing moment about the aircraft CG is:

$$N_{CG} = C_{N(CG)}qSb \qquad (16.1)$$

where:

N_{CG} = yawing moment about CG (ft-lb)
$C_{N(CG)}$ = coefficient of yawing moment about CG
q = dynamic pressure (psf)
S = wing area (ft^2)
b = wing span (ft)

rearranging equation 16.1 gives:

$$C_{N(CG)} = \frac{N_{CG}}{qSb} \qquad (16.2)$$

Because the values of q, S, and b are always positive, it follows that for a nose right (+) yawing moment, the value of the yawing coefficient must also be positive. Similarly, nose left (–) yawing moments require the value of the yawing coefficient to be negative.

Graphic Representation of Static Directional Stability

A plot of the variation of $C_{N(CG)}$ at different sideslip angles for an aircraft with positive static directional stability is shown in Figure 16.3. It has a positive slope. This is exactly opposite to the slope for static pitch stability (Figure 15.10).

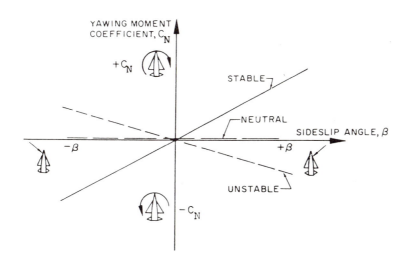

Figure 16.3   Static Directional Stability.

The trim point is where there is no yawing moment. This is where there is no sideslip angle, at the intersection of the two axes.

Assume that an airplane is at the trim point and experiences a right (+) sideslip. If the airplane has positive directional stability, it will develop a nose-right(+) yaw coefficient, which will yaw the airplane into the new relative wind. Thus, a positive slope shows positive stability.

Again, as in the case of pitch stability, the degree of the slope is an indication of the degree of the stability. Steeper slope means increased stability (or instability). It is not unusual that an airplane is stable at small sideslip angles but is unstable at high sideslip angles. This is shown in Figure 16.4. The unstable condition must not exist at angles of sideslip encountered in ordinary flight conditions.

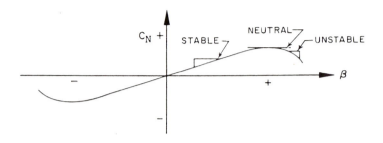

Figure 16.4   Static directional stability at high sideslip angles.

Contribution of Aircraft Components to Yaw Stability

Wings

The wing contribution to positive static directional stability is small, but it increases with the amount of sweepback. This is shown in Figure 16.5.

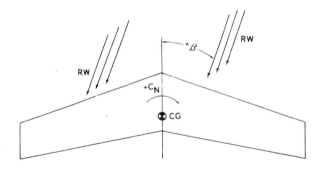

Figure 16.5   Effect of wing sweepback on directional stability.

In the above illustration the sideslip is to the right, and the right wing produces more drag than the left wing. This is a stabilizing influence. The right wing also produces more lift, but this is a roll factor and is discussed under the roll stability section of this chapter.

Fuselage

In our discussion of pitch stability we discovered that the fuselage contribution was a destabilizing factor. A similar effect occurs when we consider sideslip. The center of pressure is located near the quarter length of the fuselage, which is ahead of the CG, and is destabilizing.

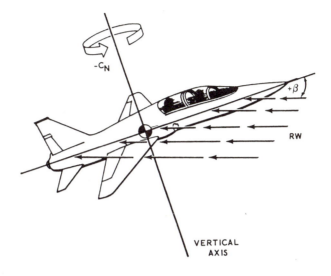

Figure 16.6   Directional instability of fuselage.

Another factor that contributes to the directional instability of the fuselage is that, if the side area of the fuselage forward of the CG is greater than that behind it, the relative wind hitting this area creates a destabilizing yawing moment. Figure 16.6 shows the fuselage effect.

Engine Nacelles

The contribution of the propeller or jet engine inlets to directional stability of an aircraft depends on their fore and aft location with respect to the aircraft's CG. If the aircraft is sideslipped, the relative wind must be changed to a direction parallel to the axis of the engine. This is similar to the discussion of pitch stability (see Figure 15.17). A sideward force is developed when the direction of the airflow is changed, and this will be a destabilizing factor if the propeller or engine inlet is forward of the CG. Aft engines are stabilizing in both pitch and yaw.

Vertical Tail

As the name implies, the vertical stabilizer is a strongly stabilizing factor. It is located behind the CG, and a sideslip creates a horizontal tail force on the aerodynamic shape of the tail. This produces a stabilizing yawing moment about the CG as shown in Figure 16.7.

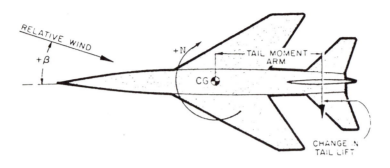

Figure 16.7   Vertical tail is stabilizing in yaw.

Tail stability can be enhanced by the addition of a dorsal fin, which does not increase the parasite drag as much as increasing the size of the vertical tail would. Another advantage of the dorsal fin is that it delays the tendency of the vertical tail to stall at high sideslip angles.

Figure 16.8 shows both the reduction of parasite drag (sketches) and the increased yaw stability (graph) resulting from the addition of a dorsal fin.

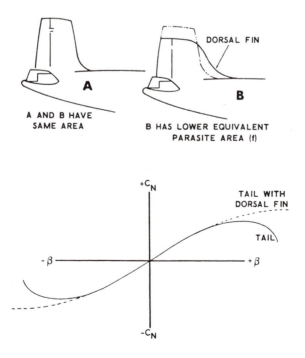

Figure 16.8   Dorsal fin decreases drag and increases stability.

Total Airplane

Figure 16.9 shows a typical build-up of the fuselage, vertical tail, and dorsal fin as they affect static directional stability. Wing and engine effects are not shown. Wing effect is small, and engine effect depends on location, as discussed earlier.

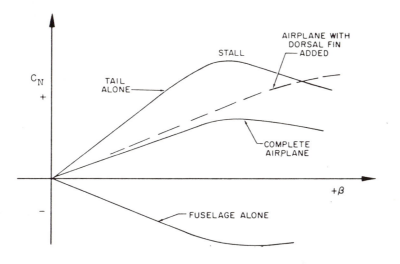

Figure 16.9   *Typical build-up of component effects on static directional stability.*

Rudder Fixed-Rudder Free Stability

Fixing the rudder in the neutral position will prevent "rudder float" and will effectively increase the vertical tail area and thus increase the directional static stability. This is similar to the stick fixed stability that we discussed under pitch stability.

For aircraft with conventional, reversible controls, increased directional stability will result if the pilot keeps his feet on the pedals and holds the rudder in the neutral position. Figure 16.10 shows this.

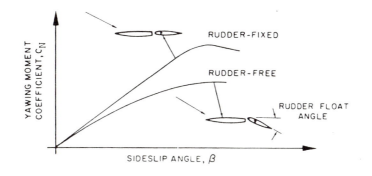

Figure 16.10   *Rudder fixed-rudder free yaw stability.*

Effect of High Angle of Attack

If the vertical tail is engulfed in stalled air from the wings at high angles of attack, it will not be effective in developing sideward forces, and static directional stability will deteriorate. This is shown in Figure 16.11. The decay in stability will have a strong effect on the ability to recover from spins and unusual attitudes.

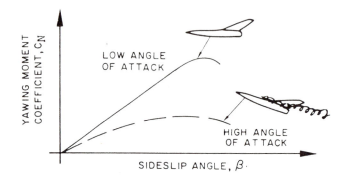

Figure 16.11   Loss of directional stability at high AOA.

DIRECTIONAL CONTROL

There are five conditions of flight that can be critical to the directional control forces exerted by the rudder. The type of airplane and its mission will determine which of these is the most important.

1. Spin Recovery

This was discussed earlier in Chapter Eleven.

2. Adverse Yaw

This is a coupled dynamic stability problem involving both directional and roll stability. It is discussed later in this chapter in the section on coupled effects.

3. Slipstream Rotation

This effect is predominant in single engine propeller aircraft. It may be critical at high power and low airspeed combinations, and thus it is important during takeoff and landing operations. The slipstream from the propeller rotates about the fuselage of U.S. aircraft as shown in Figure 16.12.

If it strikes the left side of the vertical stabilizer it will cause a nose-left yawing moment that must be overcome with rudder force to maintain directional control.

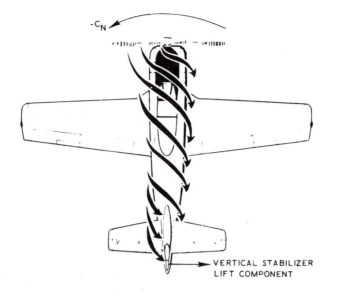

Figure 16.12   Slipstream rotation causes yaw.

4. Crosswind Takeoff And Landing

The aircraft rudder forces must be great enough to maintain a track down the runway when a cross wind exists. In effect the airplane is sideslipping with respect to the wind. The positive static yaw stability of the airplane will try to turn it into the crosswind. This is undesirable during takeoffs and landings, and the rudder must be able to overcome this tendency.

5. Asymmetrical Thrust

Multiengine airplanes have an additional control requirement that single engine airplanes do not have. If an engine malfunctions, the aircraft must be able to maintain its heading. The airplane in Figure 16.13 illustrates this requirement. The left engine is assumed to have lost thrust, and a nose-left yawing moment has resulted. To maintain the original heading an equal and opposite yawing moment must be developed by the rudder and vertical stabilizer combination.

Note that even when the moments are canceled out, there will be an unbalance in sideward forces acting on the airplane.

This will cause a sideslip to the left unless the right wing is dropped slightly. The minimum airspeed where the rudder can produce the necessary yawing moment to maintain heading, if an engine fails, is called the *in-flight minimum control speed*, V_{MCA}. The usual requirement is that V_{MCA} is not greater than $1.2 V_S$.

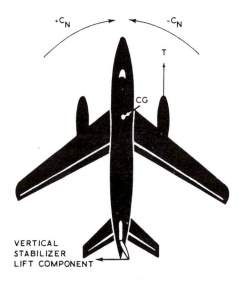

Figure 16.13   Yawing moment due to asymmetrical thrust.

LATERAL STABILITY AND CONTROL

Lateral stability and control refer to the behavior of the airplane in roll, that is, movement of the lateral axis when it is rotated about the longitudinal axis. Roll results when a rolling moment (L') acts on the airplane.

This rolling moment either can be caused by the pilot activating the ailerons (or spoilers), or it can be a result of the aircraft being subjected to a sideslip angle. From the stability standpoint, we are more interested in the reaction of the airplane to a sideslip angle.

In the next section we will discuss static lateral stability and lateral control. Dynamic lateral stability is coupled with dynamic directional stability, and they are discussed later in this chapter.

STATIC LATERAL STABILITY

Figure 16.14 shows an airplane that is sideslipping to the right. This sideslip causes a rolling moment to be developed on the aircraft. For the illustrated case the right wing will drop, and the airplane will sideslip even farther to the right. This is an unstable condition.

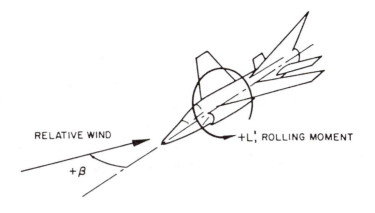

Figure 16.14   Rolling moment caused by sideslip.

For static stability in roll we need a wings leveling rolling moment to be developed if a wing drops and sideslip results. Figure 16.15 shows the three possibilities. In Figure 16.15(a) the airplane is sideslipping to its right (the viewer's left, since this view is looking aft) and is experiencing a + sideslip angle. To level the wings and have positive static roll stability, a left wing down (–) rolling moment is required. In Figure 16.15(b) there is no rolling moment developed when a right sideslip occurs, so this airplane has neutral static roll stability. Figure 16.15(c) shows an unstable airplane (the same situation shown in Figure 16.14).

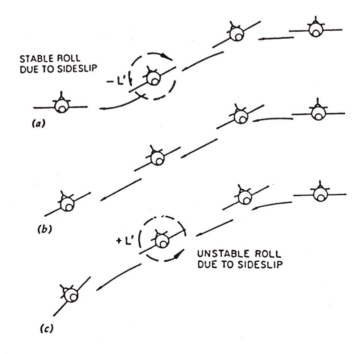

Figure 16.15   (a) Stable, (b) neutral and (c) unstable static lateral stability.

The Rolling Moment Equation

The rolling moment equation about the aircraft CG is:

$$L'_{CG} = C_{L'(CG)} Sb \qquad (16.3)$$

where:

L'_{CG} = rolling moment about the CG (ft-lb)
$C_{L'(CG)}$ = coefficient of rolling moment about CG
q = dynamic pressure (psf)
S = wing area (ft^2)
b = wing span (ft)

Rearranging equation 16.3:

$$C_{L'(CG)} = \frac{L'_{CG}}{qSb} \qquad (16.4)$$

For a right wing down, +, rolling moment, $C_{L'}$ must be positive; and for a left wing down, –, rolling moment, $C_{L'}$ must be negative.

Graphic Representation of Static Lateral Stability

A plot of the variation of $C_{L'(CG)}$ at different sideslip angles for an aircraft with positive static lateral stability is shown in Figure 16.16. The plot has a negative slope.

This slope is the same as that for static longitudinal stability, but it is opposite to that for static directional stability (see Figures 15.9 and 16.3).

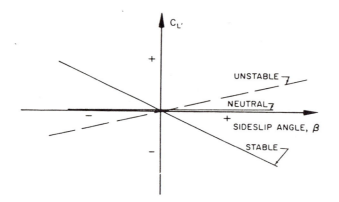

Figure 16.16   Static lateral stability.

The trim point is the point where there is no rolling moment. This occurs where there is no sideslip angle, at the intersection of the two axes.

Assume that an airplane is trimmed for zero roll and experiences a right, +, sideslip. If the airplane has positive lateral stability it will develop a negative rolling moment coefficient, $-C_{L'}$, which raises the right wing. This is stabilizing. Thus, a negative slopeshows positive stability.

As was the case with both pitch stability and yaw stability, the degree of the slope is an indication of the degree of the stability. Steep slope means a higher degree of stability (or instability).

Contributions of Aircraft Components to Roll Stability

Wing Dihedral

Dihedral is defined as the spanwise inclination of a wing or horizontal stabilizer to the horizontal, or to a plane equivalent to the horizontal. Upward inclination is positive dihedral, and downward inclination is negative dihedral, commonly called *anhedral.* Dihedral is shown in Figure 16.17.

DIHEDRAL ANGLE

Figure 16.17   Dihedral angle.

It is difficult to illustrate in a two-dimensional drawing the relative wind as it hits the aircraft in a sideslip. Figure 16.18 is an attempt to do this.

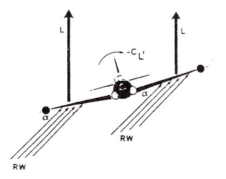

Figure 16.18   Dihedral producing static lateral stability.

The airplane in Figure 16.18 is in a sideslip to the right. The sideslip angle is positive as defined previously.

The AOA of the right wing is greater than that of the left wing because of the dihedral of the wings. Therefore, more lift L is generated on the right wing, and the negative coefficient of rolling moment, $-C_{L'}$, is generated and a stabilizing rolling moment results.

Vertical Wing Position

A high wing position places the CG of an airplane below the center of pressure of the wing. This results in a pendulum effect that is equivalent to 1 to 3° of positive effective dihedral. Conversely, a low mounted wing with the CG of the airplane above the center of pressure of the wing will be unstable and is equivalent to 1 to 3° of negative effective dihedral. This is apparent in observing the larger dihedral angles of low winged airplanes.

Wing Sweepback

The contribution of swept wings in producing positive static directional stability was discussed previously (see Figure 16.5). We saw that if a swept wing airplane was sideslipped to the right that the wing on the right had more drag than the wing on the left. The right wing also has more lift than the left wing. The relative wind hits the right wing at a more favorable angle.

This is shown in Figure 16.19. This is stabilizing, much in the same way that dihedral is stabilizing: in fact, it is called the *dihedral effect of sweepback*.

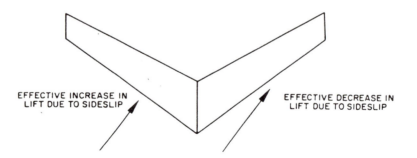

Figure 16.19   Dihedral effect of sweepback.

Vertical Tail

Side forces on the vertical tail will be a stabilizing influence because the tail is above the CG. The effect of the vertical tail in producing positive static lateral stability is shown in Figure 16.20.

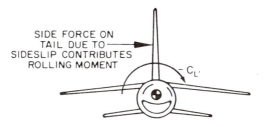

Figure 16.20   Vertical tail effect on lateral stability.

Complete Airplane

The complete airplane must have a positive lateral stability. Some components may have negative stability, but this must be overcome by the stabilizing moments of other components, and the entire aircraft must be laterally stable.

LATERAL CONTROL

The lateral control of an airplane is accomplished by providing differential lift on the wings. This is usually obtained by some type of ailerons or spoilers. Delta wing aircraft often combine the ailerons and elevators into a single control unit called an *elevon* or *ailevator*.

Both the left and right surface act together when elevator action is called for. They act in opposition to each otherwhen roll motion is required. A combination of pitch and roll response is also possible. Pilot inputs are similar to those for conventional controls.

Critical lateral control conditions are takeoffs and landings in crosswind situations. Military fighters and acrobatic airplanes may need high roll rates and therefore may desire less lateral stability and more lateral control than transport aircraft.

DYNAMIC DIRECTIONAL AND LATERAL COUPLED EFFECTS

In our previous discussion of static directional and lateral stability, we saw that the static stability depended on the aircraft's reaction to an imposed sideslip angle. Both yawing and rolling of an aircraft produce sideslip; or, conversely, sideslip produces both yawing and rolling moments. These two moments interact and result in coupled effects that determine the dynamic stability of an aircraft in yaw and roll.

Several effects of coupled yaw and roll stability are described below.

257

Roll Due to Yawing

The normal way to produce rolling moments is by use of ailerons. However, yawing can also produce roll. If the pilot applies right rudder, the aircraft yaws to the right. In rotating about the CG, to the right, the left wing moves faster than the right wing. The left wing then develops more lift, and the aircraft rolls to the right. This also ties into our previous discussion on static roll stability. The yaw to the right creates a negative sideslip angle (relative wind from the left). This causes a positive rolling moment (right wing down), and the airplane rolls to the right.

Adverse Yaw

Normally an airplane will yaw in the same direction that it is rolled. However, it is possible that an airplane will yaw in the direction opposite to the roll. This can lead to a loss of control. It is called *adverse yaw*.

First, let us examine the lift forces on each wing when an airplane is rolled. Figure 16.21 shows the rear view of an airplane that is making an aileron roll to the left.

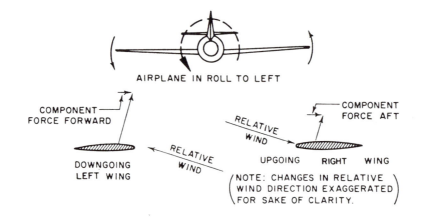

Figure 16.21   Adverse yaw.

The effective relative wind on the upgoing (right) wing is the vector resultant of the free stream relative wind and the downward relative wind, due to the wing moving upward. The lift vector is perpendicular to the effective relative wind and is thus tilted backward. This is similar to the discussion of induced drag in Chapter Five (see Figure 5.10).

The effective relative wind on the downgoing (left) wing is the vector resultant of the free stream relative wind and the upward relative wind, due to the wing moving downward. The lift vector is perpendicular to the effective relative wind and, for this wing, is tilted forward.

The rearward direction of the lift vector on the upgoing wing and the forward direction of the lift vector on the downgoing wing both oppose the yaw of the airplane in the intended direction of turn. In this example they try to turn the airplane to the right, hence the term adverse yaw.

Adverse yaw effects are amplified at high angles of attack. This is explained by referring to Figure 16.22.

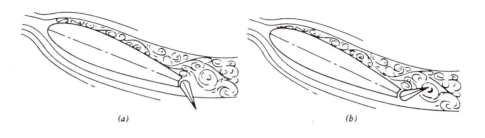

Figure 16.22   High AOA (a) upgoing wing; (b) downgoing wing.

The drag of the upgoing wing (Figure 16.22(a)) is much greater than on the downgoing wing (Figure 16.22(b)), thus additional adverse yaw tendency exists.

To decrease adverse yaw tendencies, neutralize the ailerons, once bank has been established.

Types of Motion Resulting from Coupled Effects

Three types of airplane motion can result from the interaction of yaw and roll:

1. *Spiral divergence* results when the static directional stability is great in comparison to the static lateral stability (dihedral effect). If a wing is lowered the directional stability is greater than the roll stability, and the aircraft will not sideslip readily. Thus the dihedral effect is weak, and the wing will not raise to the level position. The airplane tends to enter an ever-tightening spiral dive commonly called a *graveyard spiral.*

2. *Directional divergence* results from a negative directional stability. The airplane develops sideslip after being disturbed either in roll or yaw and develops a yawing moment moment that causes the airplane to yaw farther in the same direction. Once the yawing motion has started it will continue until the airplane is broadside to the relative wind. Obviously this condition cannot be tolerated, and design to prevent it is a prime factor in dynamic stability.

3. *Dutch roll* is a coupled directional and lateral oscillation that occurs when the dihedral effect is very strong in comparison to the directional stability.

If a stable airplane is in a sideslip to the right, for instance, it will yaw to the right. At the same time, however, the right wing will develop more lift and the airplane will roll to the left.

If not controlled, the upgoing right wing will cause sideslip to the left. The whole process will then be repeated on the left side of the airplane. When static directional stability is strong, the Dutch roll is heavily damped, but this does lead to spiral divergence. A small amount of spiral divergence is tolerated because it is greatly preferred to Dutch roll.

The three possible flight paths due to coupled dynamic directional and lateral stability are shown in Figure 16.23.

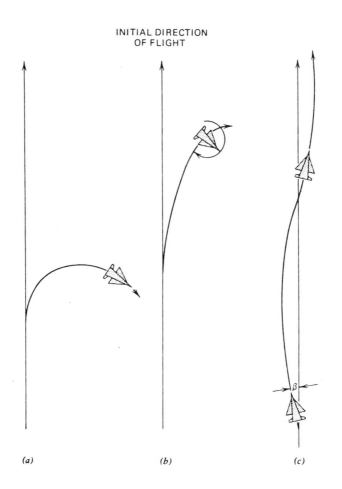

Figure 16.23 Flight paths due to coupled dynamic effects:
(a) spiral divergence;
(b) directional divergence;
(c) Dutch roll.

SYMBOLS AND UNITS

$C_{L'(CG)}$	Rolling moment coefficient about CG
$C_{N(CG)}$	Yawing moment coefficient about CG
L'_{CG}	Rolling moment about CG (ft-lb)
N_{CG}	Yawing moment about CG
β (beta)	Sideslip angle (degrees)

EQUATIONS

16.1 $N_{CG} = C_{N(CG)} qSb$

16.2 $C_{N(CG)} = \dfrac{N_{CG}}{qSb}$

16.3 $L'_{CG} = C_{L'(CG)} qSb$

16.4 $C_{L'(CG)} = \dfrac{L'_{CG}}{qSb}$

PROBLEMS

1. For static directional stability, if an airplane sideslips to the left, the plane must:

    a. keep its old heading.
    b. yaw to the left.
    c. yaw to the right.

2. The wings of an airplane contribute to static directional stability:

    a. if they are forward of the CG.
    b. if they are behind the CG.
    c. either of the above.

3. The fuselage is unstable in both static pitch and static yaw stability.

    a. True.
    b. False.

4. Engine nacelles with propellers or jet inlets behind the airplane's CG are stable in both static pitch and static yaw stability.

    a. True.
    b. False.

5. The pilot of an airplane without powered controls can increase the stability of his airplane by:

    a. resting his feet on the rudders, even though the plane is trimmed.
    b. holding the stick (or yoke) in the neutral position.
    c. both of the above.
    d. neither a. or b. above.

6. Operating an airplane at a high AOA will result in a loss of yaw stability.

    a. True.
    b. False.

7. Which of these vertical wing locations will require the most wing dihedral for static roll stability?

    a. High wing.
    b. Mid wing.
    c. Low wing.

8. Swept wings affect static roll stability more than straight wings by:

    a. increasing the stability.
    b. decreasing the stability.
    c. they do not change the stability at all.

9. If the pilot of a stable airplane applies right rudder, the airplane will:

    a. roll to the left and yaw to the left.
    b. roll to the right and yaw to the right.
    c. roll to the left and yaw to the right.
    d. roll to the right and yaw to the left.

10. Of the three types of coupled effects discussed, which is the most dangerous?

    a. spiral divergence.
    b. directional divergence.
    c. Dutch roll.

CHAPTER SEVENTEEN

HIGH SPEED FLIGHT

Flight speeds have been arbitrarily named as:

Subsonic. Aircraft speeds where the airflow around the aircraft is completely below the speed of sound (about Mach 0.7 or less).

Transonic. Aircraft speeds where the airflow around the aircraft is partially subsonic and partially supersonic (from about Mach 0.7 to Mach 1.3).

Supersonic. Aircraft speeds where the airflow around the aircraft is completely above the speed of sound but below hypersonic airspeed (from about Mach 1.3 to Mach 5.0).

Hypersonic. Aircraft speeds above Mach 5.0.

In this chapter we will discuss the airflow as the aircraft approaches the speed of sound, transonic flight, and supersonic flight. Hypersonic flight will not be discussed.

In subsonic flight, the density change in the airflow is so small that it can be neglected in the flow equations without appreciable error. The airflow at these lower speeds can be compared to the flow of water and is called *incompressible flow*. At high speeds, however, density changes take place in the airstream that are significant. Thus this type of airflow is called *compressible flow*.

Transonic, supersonic, and hypersonic flight all involve compressible flow.

THE SPEED OF SOUND

The speed of sound is an important factor in the study of high speed flight. Small pressure disturbances are caused by all parts of an aircraft as it moves through the air. These disturbances move outward from their source through the air at the speed of sound.

A two dimensional analogy is that of the ripples on a pond that result when a stone is thrown in the water.

The speed of sound in air is a function of temperature alone:

$$a = a_o \sqrt{\theta}$$

(17.1)

where
   a = speed of sound
   a_o = speed of sound at sea level, standard day
   θ = temperature ratio, T/T_o

Because the aircraft's speed in relation to the speed of sound is so important in high speed flight, airspeeds are usually measured as *Mach number* (named after the Austrian physicist Ernst Mach). Mach number is the aircraft's true airspeed divided by the speed of sound:

$$M = \frac{V}{a}$$
(17.2)

where
   M = Mach number
   V = true airspeed (knots)
   a = speed of sound (knots)

When an aircraft is flying below the speed of sound, the pressure disturbances will be moving faster than the airplane, and those disturbances that travel ahead of the aircraft influence the approaching airflow. This "pressure warning" can be observed in a smoke wind tunnel as it causes the upwash well ahead of the wing. This is illustrated in Figure 17.1(a).

If the aircraft is flying at a speed greater than the speed of sound, the airflow ahead of the aircraft is not influenced by the pressure field of the aircraft, since the speed of the pressure disturbances is less than the speed of the aircraft. The airflow ahead of the wing will have no warning of the approach of the wing and will not change its direction ahead of the wing's leading edge. This is illustrated in Figure 17.1(b).

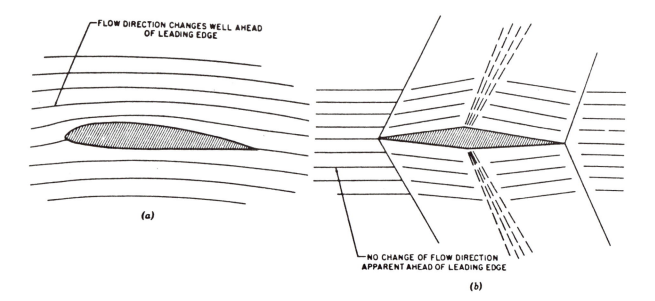

Figure 17.1   (a) Subsonic and (b) supersonic flow.

HIGH SUBSONIC FLIGHT

It is important to note that compressibility effects are not limited to aircraft that fly supersonically. A high subsonic velocity will produce local supersonic flow on top of the wings, fuselage, and other parts of the aircraft. Helicopters often experience compressibility effects on rotor tips. This can occur even when the helicopter is in a hover. The most important compressibility effect that occurs in subsonic aircraft is the formation of *normal shock waves* on the aircraft's wings or rotor blades.

Normal Shock Wave Formation on Wings

Airflow increases in velocity as it passes over a wing surface, so the local Mach number on top of the wing is always greater than the flight Mach number. For example, if an aircraft is flying at M = 0.50, the local velocity might be M = 0.78 at the thickest point of the wing (depending on the thickness and/or camber of the airfoil). This is shown in Figure 17.2.

If the flight Mach number increases, so does the local velocity on top of the wing. At some flight Mach number the local maximum velocity reaches sonic speed, M = 1.0. This flight Mach number is called the *critical Mach number*, M_{CRIT}. For the example shown in Figure 17.3, M_{CRIT} is 0.72.

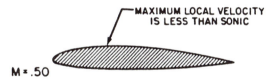

Figure 17.2   Subsonic flow about a wing section.

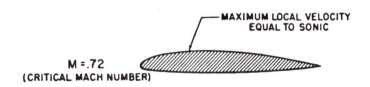

Figure 17.3   Critical Mach number.

Once M_{CRIT} is exceeded, the aircraft is flying in the tran-sonic speed range. Supersonic airflow exists in the area of maximum thickness on top of the wing; subsonic flow exists elsewhere. All pressure disturbances behind this sonic flow cannot be propagated forward because they run into sonic velocities traveling rearward. A normal shock wave is formed as shown in Figure 17.4.

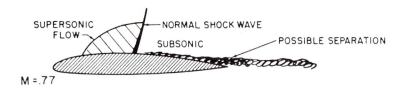

Figure 17.4   Normal shock wave formation on wing.

This shock wave is present where the air slows from supersonic to subsonic. As the air flows through the normal shock wave it undergoes a rapid compression. The compression decreases the kinetic energy of the airstream and converts it into a pressure and temperature increase behind the shock wave. The heat rise behind the shock wave is either radiated to the atmosphere or absorbed by the wing surface, but in either case it is lost, and this lost energy must be continuously supplied by the engines. this energy loss represents a type of drag known as *wave drag*.

The increase in static pressure caused by the normal shock wave has the same effect as the adverse pressure gradient that was discussed in basic aerodynamics. It slows the air in the boundary layer down to the point where faster moving air from outside the boundary layer rushes in and reverse flow occurs. This process is shown at the "possible separation" area shown in Figure 17.4.

DESIGN FEATURES FOR HIGH SUBSONIC FLIGHT

Subsonic jet aircraft can increase their airspeed without encountering shock wave problems if the critical Mach number can be raised. Several design features help accomplish this. They include:

1. Use of thin airfoil sections.

2. Use of airfoil sections that have good high subsonic MACH number characteristics.

3. Use of sweepback.

4. Use of vortex generators.

1. Use of Thin Airfoil Sections

The thinner the airfoil, the less the air is speeded up in passing over the top surface and therefore the higher the airspeed can be before the air reaches sonic speed. Thin wings do not attain high values of $C_{L(MAX)}$, so higher takeoff and landing speeds are required for thin winged aircraft.

Another disadvantage is that it is difficult to design the structural strength and rigidity required in a thin wing.

Finally, there is less room for fuel tanks in a thin wing as compared to a thicker wing of the same planform.

2. Use of High Speed Subsonic Airfoils

The laminar flow airfoil discussed in Chapter 5 was the first high speed airfoil. Moving the maximum thickness back-ward from about 25%C to 40-50%C did reduce drag and increase the critical Mach number, but one objectionable disadvantage still existed for this type of airfoil.

The shock wave that develops on a laminar airfoil occurs in the adverse pressure gradient region. As the airflow is slowing up in this region it has a tendency to separate more easily.

The *supercritical airfoil* shown in Figure 17.5 is designed to correct this deficiency. It is shaped so that the normal shock will occur where the upper surface pressure gradient is favorable or zero. At that point the boundary layer is better able to encounter the pressure increase across the shock wave without separating.

The surface curvature on the top surface of the supercritical airfoil is less than a conventional laminar airfoil, and so the local Mach number will be less than that for the laminar flow airfoil.

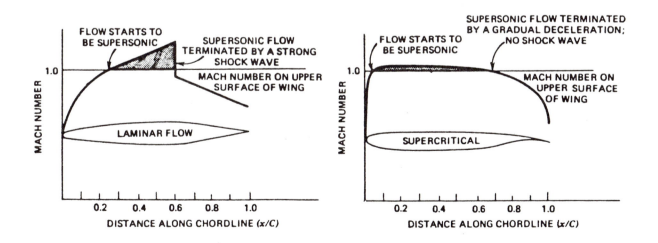

Figure 17.5   Comparison of supercritical and laminar flow airfoils at Mach 0.75.

If M_{CRIT} is exceeded, the airflow will be supersonic nearer the leading edge on the supercritical airfoil, but will remain at a much lower Mach number, and the supersonic flow will be terminated by a gradual deceleration. Therefore, no shock wave will result, and drag will be less.

3. Use Of Sweepback

One of the most common methods of increasing critical Mach number is to sweep the wings backward (and more recently, to sweep them forward). Instead of presenting the complicated conventional explanation of how this works, a much simpler description will be presented here.

We learned earlier that thin wings increase the M_{CRIT}. By sweeping the wings, the effective chord (parallel to the aircraft's longitudinal axis) has been lengthened, but the wing thickness has not been changed. Thus the ratio of thickness/chord has been reduced. In effect, the wing thickness has been reduced and a higher M_{CRIT} results.

The coefficient of lift is affected by sweep in two ways. First, the value of $C_{L(MAX)}$ is reduced, and thus takeoff and landing speeds must be increased. Second, the curve flattens out with no sharp stall AOA. Figure 17.6 illustrates these.

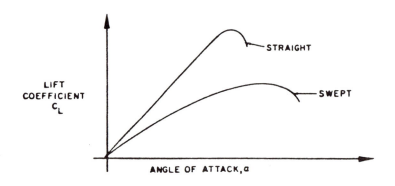

Figure 17.6   Effect of wing sweep on C_L.

The lack of a definite stall point has caused some complacency in pilots who are making a transition from straight wing to swept wing aircraft. While it is true that the swept wing aircraft does not show a clean stall, it is also true that sweeping the wings decreases the aspect ratio, and this means a large increase in induced drag coefficient. This is particularly dangerous during takeoffs and landings, as was emphasized earlier.

Wing sweep is also important in directional and lateral stability as was discussed in Chapter 16. Wing tip stall is increased by swept back wings as was presented in Chapter 11. Sweeping the wings forward will eliminate this problem.

4. Use Of Vortex Generators

Use of vortex generators to invigorate the boundary layer and thus delay separation in the low speed region of flight was discussed in Chapter 4. The same general principles apply to the high speed region of flight.

The shock-induced separation occurs because the boundary layer does not have enough energy to overcome the adverse pressure gradient through the shock wave. Here again the vortex generators will mix higher kinetic energy air from outside the boundary layer with the slowing air in the boundary layer and delay the separation process.

If a normal shock wave does develop, vortex generators are effective in breaking it up. The additional drag of the generators is small in comparison to the wave drag that they help dissipate. Figure 17.7 shows the action of vortex generators.

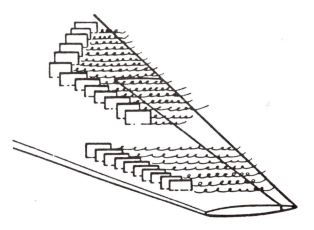

Figure 17.7   Vortex generators.

TRANSONIC FLIGHT

Early attempts to "break the sound barrier" were not very successful because of problems encountered in the transonic region. When these problems were better understood, several important design changes were made, and today supersonic aircraft have little difficulty in negotiating this speed region.

Subsonic aircraft do not have these design changes and will experience difficulties if the critical Mach number is exceeded.

This discussion, therefore, applies to subsonic aircraft and helicopters, if they venture into the transonic region of flight, as well as to supersonic aircraft.

Force Divergence

At airspeeds of about 5% above M_{CRIT} the normal shock wave on the top of the wing causes the boundary layer to separate from the wing. This causes a change in the aerodynamic force coefficients, both C_L and C_D. This airspeed is called the *force divergence Mach number.*

Figure 17.8 shows the C_D curve (for a constant value of C_L) plotted against Mach number. When the force divergence Mach number is reached, the value of C_D rapidly increases.

This velocity is also known as the *drag divergence* or *drag rise Mach number.* The drag increase is caused by the energy loss across the normal shock wave and the boundary layer separation and is called wave drag.

It must be noted that even though the value of C_D again decreases above Mach 1, that the total drag continues to increase with airspeed. The rate of increase will be less, however, above Mach 1.

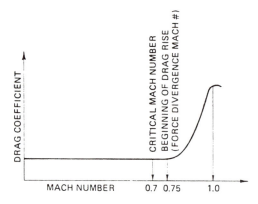

Figure 17.8   Force divergence effect on C_D.

In addition to the drag increase behind the normal shock wave, the airflow separation also results in a loss of lift. Figure 17.9 shows a sudden lowering of the coefficient of lift at the force divergence Mach number.

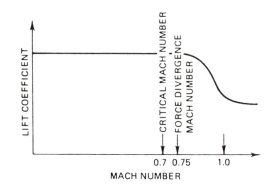

Figure 17.9   Force divergence effect on C_L.

Shock-induced separation creates a local stall situation similar to the low speed stall, except that it occurs behind the normal shock wave rather than at the trailing edge of the wing. This causes the center of pressure to be shifted forward and produces a nose-up pitching moment. If one wing develops a normal shock before the other, a rolling moment can be produced toward the wing with the earlier shock. If a wing drops, the aircraft will tend to yaw in that direction, and a condition similar to "Dutch roll" may develop.

Some of the effects of reaching force divergence Mach number are summarized as follows:

1. An increase in C_D for a given value of C_L.

2. A decrease in C_L for a given AOA.

3. A change in pitching moment as the aerodynamic center shifts.

Tuckunder

The downwash behind the wing will be decreased when airflow separation takes place. This will result in the horizontal stabilizer AOA being effectively increased, and it will develop more lift. This is one of the factors that causes the aircraft to pitch nose down; it is commonly called *tuckunder*.

If the flight Mach number is increased beyond the force divergence Mach number, the normal shock wave on top of the wing increases in intensity and moves backward. A second normal shock wave then appears on the bottom of the wing. This is shown in Figure 17.10.

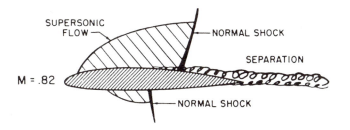

Figure 17.10 Normal shock wave on bottom of wing.

Another factor influencing tuckunder is the movement of the shock waves toward the rear of the wing. As the top shock moves rearward the separation point also moves rearward and so does the center of pressure, thus adding to the tuckunder tendency.

A third factor is that the aerodynamic center moves from the quarter chord point to the 50% chord point as the aircraft reaches supersonic flight. All aircraft flying supersonically, therefore, suffer a nose-down pitching moment.

The shift of the aerodynamic center is not a smooth movement. Sharp leading edges tend to make the shift smoother, but in many aircraft the aerodynamic center moves forward before the eventual rearward shift. This is shown in Figure 17.11. This movement can cause a pitch-up moment in the early stages of transonic flight, but eventually it will cause tuckunder as the speed increases.

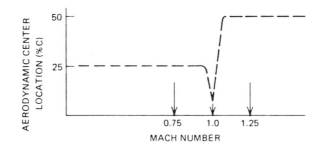

Figure 17.11   Aerodynamic center location shift.

Buffet

The violently turbulent separated air behind a normal shock wave often produces buffeting of the aircraft. This is most commonly caused by the airflow hitting the horizontal stabilizers. The T-tail configuration helps eliminate this problem by putting the horizontal tail above the wing wake. Flight in the transonic range is generally undesirable, so supersonic aircraft usually fly through this region as rapidly as possible.

Control Surface Buzz

Boundary layer separation acting on control surfaces often causes rapid oscillations called *buzz*. This can cause metal fatigue problems to hinge fittings and other parts of the control surfaces.

Control Effectiveness

Shock-induced separation can reduce the effectiveness of control surfaces in two ways. First, the surface that is located in a region of stalled air is operating in an aerodynamically dead air mass and therefore cannot produce effective aerodynamic forces.

Second, control surfaces are effective if they change the airflow around the entire wing or stabilizer. When there is a shock wave ahead of the control surface, deflection of that surface cannot influence the airflow ahead of the shock wave, and thus what little aerodynamic force is developed is restricted to the control surface and area behind the shock wave.

SUPERSONIC FLIGHT

As the flight Mach number increases in the Transonic speed range, the upper and lower normal shock waves increase in size and strength and move to the trailing edge of the wing, as shown in Figure 17.12.

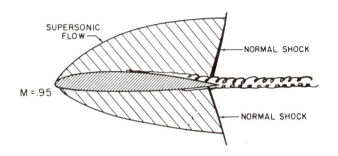

Figure 17.12   Normal shock waves move to trailing edge.

When the flight Mach number is 1.0, the entire airfoil is supersonic, except the leading edge stagnation area. The shock waves at the trailing edge are still normal shock waves, so the velocity behind them is still subsonic. A new shock wave now appears ahead of the airfoil. It is a compression wave just like the normal shock wave, and because the air at the leading edge is subsonic, the new wave, called a *bow wave*, must also be a normal shock wave, at least in the area directly ahead of the leading edge. Figure 17.13 shows a bow wave ahead of the leading edge.

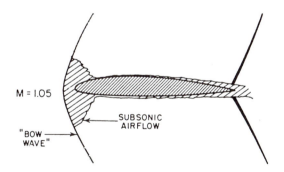

Figure 17.13   Unattached bow wave at transonic speed.

Fully supersonic flight exists when the bow wave becomes attached to the leading edge of the wings, nose of the aircraft, tail sections, and all other parts of the aircraft. The bow wave may not attach easily if the leading edges are blunt. Sharp leading edges are used on supersonic aircraft, but it still is necessary for the flight Mach number to be greater than 1.0 for **all** of the airflow to be supersonic.

Thus the supersonic range starts at about Mach 1.2 to 1.3, depending on individual aircraft design. The normal waves at the trailing edges also are deflected backwards and are now called *oblique shock waves*.

Oblique Shock Waves

An oblique shock wave, like a normal shock wave is a compression wave, but it differs from the normal shock in that the airflow direction changes as air flows through it and the airflow slows down but remains supersonic. To see how an oblique shock wave forms, consider a wedge shaped object placed in a supersonic flow as shown in Figure 17.14.

The airflow next to the horizontal surface must change direction when the wedge is encountered. It does this and assumes the direction of the arrow V_2. This sudden change in direction, as the air particles turn into the path of other particles, (V_1), traveling in the horizontal direction, produces a shock wave at an oblique angle to the horizontal. The angle that the wave makes with the horizontal is called the *wave angle*. It is also shown in Figure 17.14.

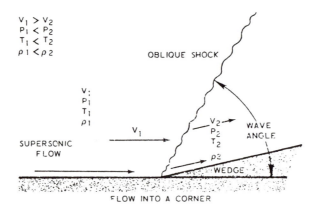

Figure 17.14 Formation of an oblique shock wave.

The wave angle depends upon the Mach number of the approaching flow and the angle of the wedge. It is the sum of the *Mach wave angle*, μ, and the *wedge angle*. The Mach wave angle is the angle whose sine is $\frac{1}{M}$ (arc sin $\frac{1}{M}$).

$$\sin \mu = \frac{1}{M}$$

(17.3)

For example, if M = 2.0 and the wedge angle is 10°, then

$$\text{wave angle} = \text{arc sin } 0.5 + 10° = 30° + 10° = 40°$$

The oblique shock wave is weaker than the normal shock, but it still is a shock wave and is wasteful of energy. As air passes through an oblique shock wave, its density, pressure, and temperature all rise, and its velocity decreases. The main difference is that the air remains supersonic behind an oblique shock wave, but it is always slowed to subsonic behind a normal shock wave.

Expansion Waves

The third kind of wave that occurs in supersonic flow is the *expansion wave*. Unlike the normal and oblique shock waves, the expansion wave is not a shock wave. In fact, the expansion wave is just the opposite of a compression wave. Density, pressure, and temperature all decrease as the air flows through an expansion wave. Air velocity increases through this wave, and no energy is lost. Expansion waves form where the airstream turn around a convex corner. This is shown in Figure 17.15.

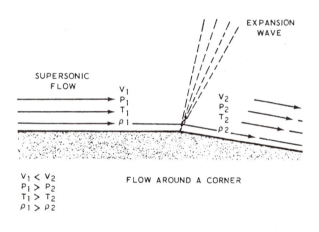

Figure 17.15 Formation of an expansion wave.

In flowing around the convex corner, the airstream makes an infinite number of minute direction changes until it has turned the corner and the flow direction is parallel to the downstream surface.

The expansion wave takes place through a fan shaped area as shown by the dashed lines in Figure 17.15. Shock waves, on the other hand, take place through a very narrow area.

As the static pressure decreases through an expansion wave, lift is produced in this region.

Subsonic flow could not negotiate a convex corner as the flow would separate from the wing. Supersonic flow, however, has a higher energy level and can turn these sharp corners by means of an expansion wave, without separation.

Characteristics of the waves are summarized below:

Type of wave formation	Oblique shock wave	Normal shock wave	Expansion wave
Flow direction change:	"Flow into a corner," turned to flow parallel to the surface	No change	"Flow around a corner," turned to flow parallel to the surface
Effect on velocity and Mach number:	Decreased but still supersonic	Decreased to subsonic	Increased to higher supersonic
Effect on static pressure and density:	Increase	Great increase	Decrease
Effect on energy or total pressure:	Decrease	Great decrease	No change (no shock)
Effect on temperature:	Increase	Great increase	Decrease

Figure 17.16   Supersonic wave characteristics summary.

Aerodynamic Forces in Supersonic Flight

Aerodynamic forces are produced by differential pressures acting on the surfaces of a wing in an airstream. This is true whether the airstream is subsonic or supersonic. In supersonic flight sudden increases in pressure result when the airflow passes through shock waves. Expansion waves produce less rapid changes in pressure. The pressure drops when passing through an expansion wave. An expansion wave at a sharp corner produces a fairly rapid pressure drop, while an expansion over a curved surface will produce a gradual pressure drop.

Supersonic Airfoils

One type of supersonic airfoil is the *double wedge airfoil* shown in Figure 17.17. This airfoil cannot be used subsonically because the air will become separated when it reaches the convex angles at the top and bottom of the wedges. It does have use in supersonic missiles that are boosted through the subsonic flight region.

Figure 17.17(a) shows the wave pattern about a double wedge airfoil at zero AOA and thus zero lift in supersonic flight. Oblique shock waves, of the same strength, form at the top and the bottom of the leading and trailing edges. Behind these shocks the pressure rises but no net lift results. At the sharp corners at the 50% chord point expansion waves form, again of the same strength. Lower pressures exist behind the expansion waves, but again the pressures on the top and bottom are equal in value and cancel each other.

This is shown in Figure 17.17(b). The rearward components of the pressure arrows represent "wave drag" which exists even when no lift is being developed.

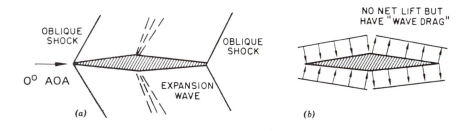

Figure 17.17   Double wedge airfoil in supersonic airflow: (a) wave pattern; (b) pressure distribution.

Now consider the double wedge airfoil at a small positive AOA, as shown in Figure 17.18(a). The oblique shock waves are now not equal in strength. The shock wave at the top of the leading edge does not have to turn as high an angle as it did at zero AOA and is now weaker. The pressure rise through the top shock is now lessened. The reverse is true for the bottom leading edge shock; it is stronger, and the pressure rise is greater than at zero AOA. The expansion waves at the midpoint corners are not of equal strength at a positive AOA. The top expansion is now greater. because the angle to be turned is greater than it was at zero AOA. The pressure decrease is also greater on the top. The opposite reasoning applies to the bottom expansion wave, and the pressure decrease behind the bottom expansion wave is lessened. Net lift is now being developed. The pressure distribution is shown in Figure 17.18(b).

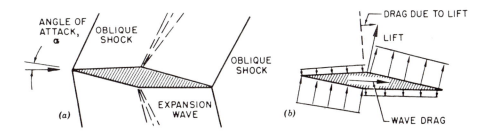

Figure 17.18   Double wedge airfoil developing lift: (a) wave pattern; (b) pressure distribution.

To avoid the subsonic problems of the *double wedge airfoil,* another supersonic airfoil section is used. This is called the circular arc or bi-convex airfoil. It uses two arcs of circles to define its shape.

This airfoil develops shock waves at the leading and trailing edges, but since there are no sharp angles at the mid-chord point, no concentrated expansion wave occurs at that point.

Instead a continuous expansion wave, from the leading edge to the trailing edge, forms on the top and bottom airfoil surfaces. A symmetrical circular arc airfoil produces no net lift at zero AOA because the waves are symmetrical. But at positive AOA the shock wave on the top leading edge is weaker and the top expansion wave is stronger than the corresponding waves on the bottom of the airfoil.

Thus positive lift is created as shown in Figure 17.19.

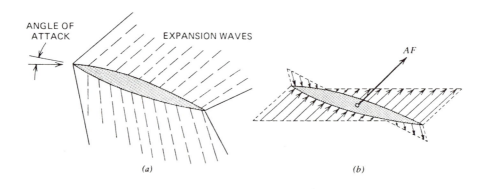

Figure 17.19   Circular arc airfoil in supersonic flow: (a) wave pattern; (b) pressure distribution.

The center of pressure and the aerodynamic center are both at the same position for both the double wedge and the circular arc airfoils. This position is at the 50% chord point.

Even though the circular arc airfoil can be flown at subsonic speeds, it does not develop a high value of $C_{L(MAX)}$ and thus takeoff and landing speeds are high.

Wing Planform

Earlier in this chapter we discussed the use of sweepback as a means of increasing M_{CRIT} for subsonic aircraft. This same principle has other advantages for supersonic aircraft as well. In addition to increasing the force divergence Mach number, sweepback also reduces the amount of C_D increase.

This is shown in Figure 17.20. It can be seen that the increase in drag coefficient is both delayed and reduced in value by sweepback. It is also interesting to compare straight wings, (0° sweep), with swept wings.

Note that the straight wing and the swept wing curves cross at some Mach number greater than 1.0. For instance, for the 45° wing this is at about M = 1.4.

This means that at this speed both wings would have the same drag and that at higher speeds the straight wing actually has less drag. Therefore, for high supersonic aircraft the straight wing configuration may be desirable. Of course at lower speeds than the crossover point, the advantage is with the swept wing aircraft.

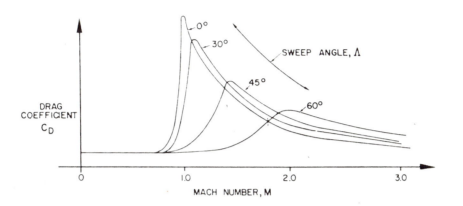

Figure 17.20   Effect of wing sweep on C_D.

In our earlier discussion of the formation of oblique shock waves we limited our analysis of Mach wave angle to two dimensional flow. But in reality the pressure disturbances radiate outward in all directions, like expanding spheres. Thus, instead of a Mach wave we have a *Mach cone*. This is shown in Figure 17.21.

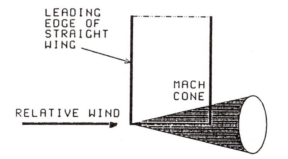

Figure 17.21   Mach cone.

In drawing illustrations it is much easier to show them in two dimensions, thus the term Mach wave is used instead of Mach cone in many of the figures.

Another advantage of sweep is that it places the wing within the Mach wave (cone) of the aircraft. This allows us to use subsonic airfoil sections for supersonic flight. Use of subsonic airfoils is desirable because they develop higher $C_{L(MAX)}$ values and thus reduce takeoff and landing speeds.

Consider a Mach wave that is generated at the wing root as shown in Figure 17.22. The air is supersonic ahead of the Mach wave and it is also supersonic in the direction of flight behind the Mach wave. The component of air behind the Mach wave that is perpendicular to the Mach wave will, however, be subsonic.

If the wing is swept back so that its leading edge is behind the Mach wave, it will be flying in a subsonic flow and thus can use a subsonic airfoil section.

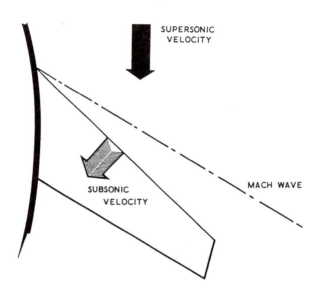

Figure 17.22   Swept wing in supersonic flight.

Area Rule Drag Reduction

Experiments in wind tunnels showed that bodies with sudden changes in cross sectional area produced more drag than similar bodies with gradual changes in area. This principle was then applied to aircraft. A large increase in the cross sectional area occurs where the wings attach to the fuselage. By reducing the fuselage area at this point, the total area was enlarged more gradually and drag was reduced. At the trailing edge of the wing the area of the fuselage was again increased to compensate for the loss of wing area. This results in the "coke bottle" shape of the fuselage.

Use of All Movable Control Surfaces

Early attempts to "break the sound barrier" were often disastrous because the aircraft did not have the design features found on more modern aircraft. To attain the required speed, the aircraft had to be put into a steep dive. When the aerodynamic center shifted to the 50%C location, the aircraft's center of gravity was ahead of the aerodynamic center.

This created a stable dive condition that could not be overcome with the elevators, and often the aircraft continued in the dive until it hit the earth. The elevators were constructed in the conventional manner, as shown in Figure 17.23. The control forces were limited to the small elevator itself and could not influence the flow about the horizontal stabilizer. The amount of force that the pilot could exert was also limited, because no power assist controls were available.

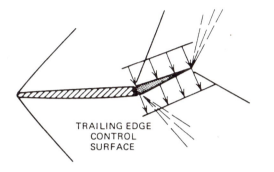

TRAILING EDGE
CONTROL
SURFACE

Figure 17.23   Subsonic control surface.

As the state of the art progressed, all moving control surfaces, as shown in Figure 17.24, were developed, as were fully powered, irreversible control systems. Much more control was then available, and pitch control was no longer a problem.

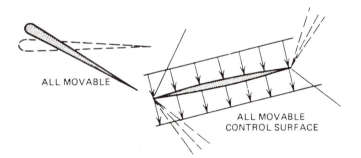

ALL MOVABLE

ALL MOVABLE
CONTROL SURFACE

Figure 17.24   Supersonic control surface.

Supersonic Engine Inlets

The compressor blades of a turbine engine can be made to operate either subsonically or supersonically, but not both. The necessity for slow speed flight, as well as supersonic flight, rules out the use of supersonic flow in the compressor. Therefore, the flow must be slowed to subsonic when the aircraft is flying supersonically. This can only be done by the airflow passing through a normal shock wave.

The Mach number behind a normal shock wave is approximately $1/M$, where M is the Mach number ahead of the shock. If the flight Mach is only slightly above 1.0 this is no problem, and a normal shock inlet such as shown in Figure 17.25 can be used.

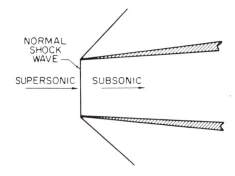

Figure 17.25   Normal shock engine inlet.

As the flight Mach number is increased, the reduction in airspeed in the normal shock wave becomes greater, the intensity of the shock increases, and the energy loss in the shock wave is increased. To reduce this energy loss an oblique shock inlet is used. This type of inlet may use a "spike" or a "ramp." These cause oblique shock waves to form.

When air passes through an oblique shock wave its velocity is reduced but remains supersonic. The energy lost in passing through an oblique shock is much less than that lost passing through a normal shock.

Often the air is slowed by a series of oblique shocks, each slowing the air, until the air velocity is only slightly supersonic; then the air is passed through a weak normal shock. Single and multiple oblique shocks formed by spike type inlets are shown in Figure 17.26.

The Mach cone angle varies with the flight Mach number, so for proper position of the shock waves, the spike (or ramp) must be adjustable in flight.

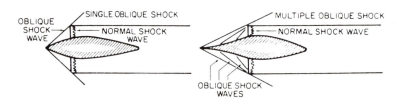

Figure 17.26   "Spike" oblique shock engine inlets.

Aerodynamic Heating

The temperature in the stagnation area of the leading edge of the wing, at the leading edges of the horizontal and vertical stabilizers, and at other stagnation points of an aircraft is shown in Figure 17.27. Strength of structural metals is shown in Figure 17.28.

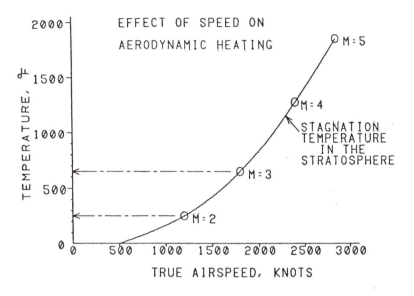

Figure 17.27 Stagnation temperatures.

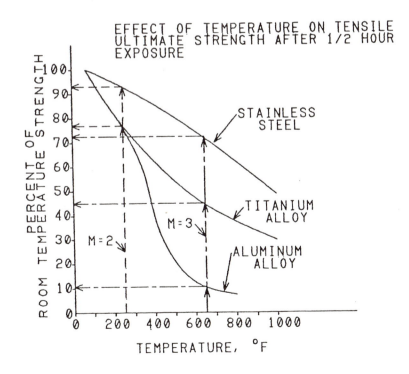

Figure 17.28 Effect of temperature on tensile ultimate strength of metals after 1/2 hr. exposure.

Figures 17.27 and 17.28 help illustrate the vast difference in problems of structural strength in designing a Mach 3 supersonic transport aircraft as compared to a Mach 2 SST. The stagnation temperature in the stratosphere is about 250° F for the Mach 2 aircraft and about 700° F for the Mach 3 SST. Comparing the tensile strength of aluminum alloy at these temperatures shows that at 250° F about 80% of the room temperature tensile strength remains, but at 700° F, less than 10% of the room temperature strength remains. Clearly, it is not feasible to make leading edges out of aluminum alloy for a Mach 3 aircraft.

SYMBOLS AND UNITS

English Symbols

a	speed of sound (knots)
a_o	speed of sound at sea level, standard day
M	Mach number (dimensionless)

Greek Symbols

θ(theta)	Temperature ratio = T/T_o
μ(mu)	Mach wave angle (degrees)

EQUATIONS

17.1 $\quad a = a_o \sqrt{\theta}$

17.2 $\quad M = \dfrac{V}{a}$

17.3 $\quad \sin \mu = \dfrac{1}{M}$

PROBLEMS

1. The speed of sound is an important factor in high speed flight because:

    a. M_{CRIT} occurs at M = 1.
    b. the pressure waves generated by the plane move at sonic speed.
    c. shock waves form when local air velocities are supersonic.
    d. both b. and c. above.
    e. none of the above.

2. The speed of sound depends on the air density, pressure and temperature.

    a. True.
    b. False.

3. Critical Mach number, M_{CRIT}, is the aircraft's speed when:

    a. it goes supersonic.
    b. the airflow first reaches sonic speed.
    c. when shock waves form.
    d. both b. and c. above.
    e. none of the above.

4. After passing through a normal shock wave, airflow is:

    a. subsonic.
    b. not changed in direction.
    c. heated up.
    d. increased in pressure and density.
    e. all of the above.

5. Airflow passing through an oblique shock wave remains supersonic but changes in direction.

    a. True.
    b. False.

6. Transonic flight problems may include:

    a. force divergence.
    b. increase in C_D.
    c. decrease in C_L.
    d. tuckunder.
    e. buffet.
    f. control surface buzz.
    g. loss of control effectiveness.
    h. all of the above.
    i. all of the above except a.

7. Above about Mach 2, the straight wing aircraft has a lower drag than a 60 degree swept wing a/c.

    a. True.
    b. False.

8. Mach wave angle changes with airspeed as follows:

    a. it decreases as airspeed increases.
    b. it increases as airspeed increases.
    c. it remains the same with airspeed changes.

9. All supersonic aircraft have circular arc or bi-convex supersonic airfoils.

    a. True.
    b. False.

10. Airflow passing through an expansion wave:

    a. speeds up.
    b. increases the energy of the airstream.
    c. decreases the temperature of the air.
    d. both a. and c. above.

ANSWERS TO PROBLEMS

Chapter One

1. d; 2. c; 3. b; 4. c; 5. d; 6. c; 7. c; 8. d; 9. d; 10. b.
11. m = 500 slugs.
12. a = 12 fps^2.
13. Drag = 282 lb, Lift = 102.6 lb.
14. Tan = 0.75, Distance = 5 mi.
15. s = 8 ft., F = 30 lb.
16. PE = 80×10^6 ft-lb, KE = 10×10^6, TE = 90×10^6 ft-lb.
17. HP = 5000.
18. F_B = 1260 lb.
19. Tension = 50 lb.
20. Tension = 6.7 lb.

Chapter Two

1. c; 2. c; 3. a; 4. a; 5. d; 6. d; 7. c; 8. b; 9. a; 10. b.
11. δ = 0.929, P.A. = 2000 ft., θ = 1.078, σ = 0.8618, D.A. = 5000 ft.
12. H = 2062.8 psf, P_2 = 1931 psf, q_2 = 131 psf, A_3 = 12.5 ft^2, P3 = 2041 psf, q_3 = 21 psf.
13. q = 116.8 psf.
14. EAS = 337.5 knots.
15. TAS = 495.5 knots

Chapter Three

1. d; 2. c; 3. d; 4. d; 5. c; 6. e; 7. d; 8. d; 9. b; 10. d.

Chapter Four

1. a; 2. c; 3. a; 4. c; 5. d; 6. d; 7. d; 8. b; 9. d.
10. a. combination, b. energy add., c. combination, d. camber chg.
11. AOA = 4°
12. V_S = 134 knots.
13. V_S = 116 knots.
14. R_E = 10.3 x 10^6

Chapter Five

1. d; 2. c; 3. d; 4. d; 5. b; 6. d; 7. d; 8. c; 9. b; 10. b.
11. D_{MIN} = 1020 lb.
12. V = 172 knots.
13. $D_i = D_p$ = 510 lb.
14.

V_2	$(V_2/V_1)^2$	D_p	$(V_1/V_2)^2$	D_i	D_T
125	0.530	270	1.89	963	1233
150	0.763	389	1.31	669	1058
172	1.0	510	1.0	510	1020
200	1.35	690	0.74	377	1067
300	3.04	1552	0.33	168	1720
400	5.42	2766	0.184	94	2860

Chapter Six

1. d; 2. d; 3. a; 4. a; 5. a; 6. a; 7. d; 8. c; 9. c; 10. b.
11. T = 3000 lb., Efficiency = 77%.
12. C_T = 1.5, T = 4100 lb., FF_0 = 6150 lb/hr.
13. FF_{TROP} = 1845 lb/hr.
14. (a) 500 kts., (b) 240 kts., (c) 0.327 (19°), (d) 10,940 fpm, (e) 240 kts., (f) 300 kts., (g) 350 kts.

Chapter Seven

1. d; 2. c; 3. a; 4. b; 5. c; 6. a; 7. d; 8. c; 9. b; 10. d.
11. V_{12} = 325 kts., V_8 = 285 kts.
12. SR_{12} = 0.2708 nmi/lb, SR_8 = 0.3455 nmi/lb.
13. SR_8 = 0.3125 nmi/lb. SR improvement in prob. 13 = 15.4%, SR improvement in prob. 12 = 27.6%. Conclusion: reduce throttle as fuel is burned.
14. SR_0 = 0.2973, SR_{20} = 0.4108, SR improvement = 38%.

Chapter Eight

1. a; 2. d; 3. c; 4. b; 5. d; 6. b; 7. b; 8. d; 9. c; 10. b.
11. 474 HP, 488, 540, 657, 1588, 3520.
12. (a) 340 kts., (b) 0.5525 (33.5°), (c) 0.32 (18.6°), (d) 4207.5 fpm, (e) 4867.5 fpm, (f) 100 kts, (g) 140 kts., (h) 160 kts.

Chapter Nine

1. c; 2. a; 3. b; 4. c; 5. c; 6. c; 7. c; 8. b; 9. b; 10. a.
11. V_{30} = 160 knots, V_{20} = 130 knots.
12. SR_{30} = 0.3765 nmi/lb, SR_{20} = 0.5778 nmi/lb.
13. SR_{20} = 0.4471 nmi/lb. Improvement = 18.7%. Improvement in prob 12 = 53.5%. Conclusion: reduce throttle as fuel is burned.
14. SR_0 = SR_{20} = .058 nmi/lb. Conclusion: no change in SR with altitude (considering airframe only).

Chapter Ten

1. c; 2. b; 3. c; 4. a; 5. d; 6. b.

Chapter Eleven

1. d; 2. b; 3. d; 4. b; 5. b; 6. a; 7. c; 8. c; 9. d; 10. d.
11. RC = 506 fpm.
12. RC = –658 fpm.

Chapter Twelve

1. d; 2. c; 3. e; 4. d; 5. a; 6. d; 7. d; 8. a; 9. a.
10. $a = 8$ fps^2.
11. $v = 126.5$ fps (75 knots).
12. $S = 2576$ ft.
13. $S = 5796$ ft.
14. $S = 2087$ ft.
15. $S = 4025$ ft.

Chapter Thirteen

1. d; 2. b; 3. d; 4. d; 5. d; 6. a; 7. a; 8. c; 9. a; 10. a.
11. $a = -8$ fps^2.
12. $S = 1785$ ft.
13. $S = 2901$ ft.
14. $S = 2160$ ft.
15. $S = 2231$ ft.

Chapter Fourteen

1. b; 2. c; 3. d; 4. d; 5. d; 6. a; 7. c; 8. d; 9. d; 10. b.
11. 30.6°.
12. Verified.
13. 1.162 G.
14. ROT = 3.23°/sec.
15. Verified.

Chapter Fifteen

1. b; 2. c; 3. d; 4. c; 5. c; 6. d; 7. a; 8. e; 9. b; 10. d.

Chapter Sixteen

1. b; 2. c; 3. a; 4. a; 5. c; 6. a; 7. c; 8. a; 9. b; 10. b .

Chapter Seventeen

1. d; 2. b; 3. b; 4. e; 5. a; 6. h; 7. a; 8. a; 9. b; 10. d.

REFERENCES

BOOKS

1 Abbott, I.H. and A.E. VonDoenhoff, *Theory of Wing Sections*, Dover, New York, 1959.

2 Anderson, J.D.Jr., *Introduction to Flight*, McGraw-Hill, New York, 1978.

3 Bent, R.D. and J.L. McKinley, *Aircraft Powerplants*, 4th ed., Gregg Division, McGraw-Hill, New York, 1978.

4 Carrol, R.L., *The Aerodynamics of Powered Flight*, Wiley, New York, 1960.

5 Dalton, S., *The Miracle of Flight*, McGraw-Hill, New York, 1977.

6 Dole, C.E., *Flight Theory and Aerodynamics*, Wiley, New York, 1981.

7 _____ *Mathematics and Physics for Aviation Personnel*, University of Southern California, 1973.

8 _____ *Fundamentals of Aircraft Material Factors*, (published by author), 1986.

9 Domasch, D.O., S.S. Sherby, and T.F. Connolly, *Airplane Aerodynamics*, 4th ed., Pitman, New York, 1967.

10 Dwinnell, J.H., *Principles of Aerodynamics*, McGraw-Hill, New York, 1949.

11 Etkin, B., *Dynamics of Atmospheric Flight*, Wiley, New York, 1972.

12 Hoerner, S.F., *Fluid-Dynamic Drag*, (published by author), 1965.

13 Hoerner, S.F. and H.V. Borst, *Fluid-Dynamic Lift*, (published by Mrs. L.A. Hoerner), 1967.

14 Kuethe, S., *Foundations of Aerodynamics*, 3rd ed., Wiley, New York, 1976.

15 Martynov, A.K., *Practical Aerodynamics*, Macmillan, New York, 1965.

16   Montgomery, J.R., *Sikorsky Helicopter Flight Theory for Pilots and Mechanics*, Sikorsky Aircraft, Stratford CT, 1964.

17   Nicolai, L.M., *Design of Airlift Vehicles*, University of Dayton, Dayton, Ohio, 1974.

18   Perkins, C.D. and R.E. Hage, *Airplane Performance, Stability and Control*, Wiley, New York, 1949.

19   Roed, A., *Flight Safety Aerodynamics*, Aerotech, Kungsangen, Sweden, 1972.

20   Roland, H.E. and J.F. Detwiler, *Fundamentals of Fixed and Rotary Wing Aerodynamics*, University of Southern California, 1967.

21   Saunders, G.H., *Dynamics of Helicopter Flight*, Wiley, New York, 1975.

22   Seckel, E., *Stability and Control of Airplanes and Helicopters*, Academic Press, New York, 1964.

23   Shapiro, A.H., *The Dynamics and Thermodynamics of Compressible Fluid Flow, Vols. I and II*, Ronald Press, New York, 1953.

24   Sutton, O.G., *The Science of Flight*, Pelican Books, (A 209), 1955.

U.S. GOVERNMENT PUBLICATIONS

25   Anonymous, *Aerodynamics*, U.S. Army Material Command, Washington, 1965.

26   Anonymous, *Aerodynamics for Pilots*, Air Training Command, U.S.A.F., ATC Manual 51-3, 1979.

27   Anonymous, *Academics for Basic*, Air Training Command, U.S.A.F., ATC Study Guide P-V4A-A-AB-SW, 1975.

28   Anonymous, *Applied Aerodynamics*, Air Training Command, U.S.A.F., ATC Study Guide P-V4A-A-AA-SW, 1977.

29   Anonymous, *Interceptor*, Air Defense Command, U.S.A.F., ADCPI 62-1 thru ADCPI 62-16, June 1964 - April 1966.

References

30   Anonymous, *Low Level Wind Shear*, Federal Aviation Administration, Advisory Circular AC No. 00-50, 1976.

31   Anonymous, *Pilot Windshear Guide*, Federal Aviation Administration, Advisory Circular AC No. 00-54, 1988.

32   Anonymous, *Rotary Wing Flight*, U.S. Army Field Manual, 1-51, 1979.

33   Anonymous, *Wake Turbulence*, Air Training Command, U.S.A.F., Pamphlet 51-12, 1976.

34   Hurt H.H. Jr., *Aerodynamics for Naval Aviators*, NAVWEPS 00-80T-80, Government Printing Office, 1965.

PERIODICALS

35   Anonymous, "Hydroplaning, Fun or Disaster," Aerospace Safety, U.S.A.F. Airforce Inspection and Safety Center, May, 1964.

36   Anonymous, "Low Altitude Wind Shear," Aerospace Safety, U.S.A.F. Airforce Inspection and Safety Center, Sept. 1976.

37   Eggleston, B. and D.L. Jones, "The Design of Lifting Supercritical Airfoils Using a Numerical Optimization Method," Canadian Aeronautics and Space Journal, Vol. 23, No 3, May/June 1977.

38   Haines P. and J. Luers, "Aerodynamic Penalties of Heavy Rain on Landing Aircraft," Journal of Aircraft, Vol. 20, No.2, Feb 1983.

39   Kocivar, B., "Super STOL," Popular Science, Feb 1976.

40   Luers J. and P. Haines, "Heavy Rain Influence on Airplane Accidents," Journal of Aircraft, Vol. 20, No 2, 1983.

INDEX